Complex Harmonic Splines, Periodic Quasi-Wavelets

Complex Harmonic Splines, Periodic Quasi-Wavelets

Theory and Applications

by

Han-lin Chen

Institute of Mathematics,
Academia Sinica,
Beijing, P.R. China

SPRINGER SCIENCE+BUSINESS MEDIA, B.V.

A C.I.P. Catalogue record for this book is available from the Library of Congress.

ISBN 978-94-010-5843-8   ISBN 978-94-011-4251-9 (eBook)
DOI 10.1007/978-94-011-4251-9

Printed on acid-free paper

Contents

Preface

This book, written by our distinguished colleague and friend, Professor Han-Lin Chen of the Institute of Mathematics, Academia Sinica, Beijing, presents, for the first time in book form, his extensive work on complex harmonic splines with applications to wavelet analysis and the numerical solution of boundary integral equations. Professor Chen has worked in Approximation Theory and Computational Mathematics for over forty years. His scientific contributions are rich in variety and content. Through his publications and his many excellent Ph.D. students he has taken a leadership role in the development of these fields within China. This new book is yet another important addition to Professor Chen's quality research in Computational Mathematics.

In the last several decades, the theory of spline functions and their applications have greatly influenced numerous fields of applied mathematics, most notably, computational mathematics, wavelet analysis and geometric modeling. Many books and monographs have been published studying real variable spline functions with a focus on their algebraic, analytic and computational properties. In contrast, this book is the first to present the theory of complex harmonic spline functions and their relation to wavelet analysis with applications to the solution of partial differential equations and boundary integral equations of the second kind. The material presented in this book is unique and interesting. It provides a detailed summary of the important research results of the author and his group and as well as others in the field.

This book is organized into four chapters. Chapter I provides a rigorous study of the functional and geometrical properties of complex harmonic spline functions. Specifically, it contains the general theory of the interpolating and quasi-interpolating complex spline functions on the boundary of the unit disc. It also contains a discussion how the boundary values of complex harmonic spline functions influence their interior behaviour. An algorithm for the computation of complex harmonic spline functions is also provided. In Chapter II various types of periodic quasi-wavelets are constructed using real and complex spline functions as generators. The orthogonality and least number of terms in the decomposition formulas for periodic quasi-wavelets, which are very important in applications, are thoroughly discussed. In Chapter III , the author applies periodic quasi-wavelets to solve boundary value problems for the two-dimensional Helmholtz equation, by reducing it to Fredholm integral equation of the second kind with a weakly singular kernel. Under certain smoothness conditions on the coefficients and the stiffness matrix being given, it is proved that the order of complexity of this algorithm is $O(N)$, where N represents the number of unknowns. In Chapter IV another type of periodic wavelets is constructed. These explicitly given wavelets possess the following important properties: interpolation, localization, symmetry, regularity up to any prescribed order, real-valued and biorthogonal. Some illustrative examples are provided.

In summary, this book is a rigorous presentation of the numerous interesting mathematical properties and physical applications of complex harmonic spline functions which is suitable not only as a reference source but also as a textbook for a special topics course or seminar. We are delighted to see the publication of this book and hope that it will foster new research and applications of complex harmonic splines and wavelets. We enthusiastically recommend it to the mathematics and engineering communities.

Charles A. Micchelli
IBM T.J.Watson Research Center
Yorktown Heights, New York, U.S.A

Yuesheng Xu
Department of Mathematics
North Dakota State University
Fargo, North Dakota, U.S.A

Introduction

In this book we first study the complex harmonic spline function (CHSF in abbreviation). We explore the functional and geometrical properties of CHSF in details. Then we apply it to some two-dimensional problems. The crucial problem in construction of CHSF is how to construct the boundary function of CHSF. We present several kinds of functions for this.

The motivation of study CHSF is from the following: The first is that the CHSF can be decomposed into elementary functions, the second is the CHSF is a good approximate method on the unit disc and the third is that the CHSF can be applied to some practical problems.

It is well known that many mathematical and physical problems can be reduced to Dirichlet problems. For instance, in real dimension 2 the Poisson problem and Neuman problem can be reduced to certain kinds of Dirichlet problems which in turn are equivalent to constructing conformal mappings (cf. [H]).

The development of conformal mappings has had a very long history, and is still full of vigour and vitality. The main reason is that it unearths many new applications in various areas of applied sciences, such as ion optics, atomic physics, nonlinear diffusion and solidification problems, etc. (cf. Chapter III, Note 1).

It is also well known that it is difficult to construct explicitly a conformal mapping which maps an arbitrary Jordan domain D', i.e., a simply connected domain bounded by a closed Jordan curve, onto another Jordan domain D, although its existence is given by Riemann mapping theorem. If we use some functions to approximate the conformal mapping, we would ask: is it a good approximant? If we obtain the map by solving certain integral equation, we like to know: how fast is it in computation? These are important problems, and they are not so easy to deal with.

In this book we will not attempt to give a full introduction to the theory of conformal mappings, since there already exist many excellent references (cf. [DG], [Gi], [Ne], [H], [LS], [Mar], [SL] and [T], etc.).

In the early 1980's we used complex harmonic spline functions (abbreviated as CHSF) to approximate complex harmonic functions to high accuracy, and each of these spline functions can be represented by a linear combination of some elementary harmonic functions (cf. [C4], [CH1] and [CH2]). Later we engaged in a systematic study of the function-theoretic and geometric properties of CHSF (see [C11]), where we extended the methods and ideas of the author's earlier works on quasi-conformal maps and complex analysis (cf. [C12]-[C15]).

We characterize CHSF in Chapter I in a series of theorems. It is clear in our discussion that CHSF reveals as a good approximant for complex harmonic functions which include conformal mapping.

The first problem in constructing the CHSF is to find an adaptive function as the boundary value of the CHSF. In Chapter I we introduce two kinds of spline functions: the interpolating complex spline functions on Γ, the boundary of the unit disc U, and the quasi-interpolant complex spline on Γ. We give a construction for the latter and provide a proof of the existence and uniqueness for the former. For the low degree complex interpolating spline functions we also give explicit expressions. How the boundary values influence the interior behavior of a CHSF is the subject to be studied in sections 3 and 4. We also provide an algorithm for computing the CHSF in section 6 of Chapter I. The approximant to the conformal mapping G from an arbitrary domain D' onto another arbitrary domain D is given by the compound function $F(D', D, \mathcal{Z})$ at the end of Chapter I.

Differing from Meyer's and Daubechies' approach on periodizing orthonormal wavelets (refer to [Myr] and [Da]) on the real axis, in Chapter II, we construct various types of periodic quasi-wavelets (abbreviated as PQW). Beginning with different spline functions (real and complex polynomials and trigonometric polynomials) as the generators, we also employ more general functions as the generators to create PQW. Besides the orthogonality, a remarkable property of PQW is that in many cases PQW has the least number of terms either in the decomposition formulas or in the reconstruction formulas of coefficients (see formulas (2.1.49), (2.1.46), (2.2.14) and (2.2.15)). Note that these formulas can be written in a tight form, for example, (2.1.41) and (2.1.42), etc. The transform matrix M_m defined by (2.1.43) is unitary. These properties make PQW very important in applications.

In Chapter III we apply PQW to solve the boundary value problems for the 2-dimensional Helmholtz equation. This equation is usually solved by boundary integral methods, which leads to Fredholm integral equation of the second kind with weak singular kernel (see $(3.0)_1$). The Dirichlet problem can also be reduced to this kind of integral equation. In particular, the exterior boundary problem of acoustic waves scattering can be reduced to such an integral equation. Thereby, integral equation $(3.0)_1$ has a strong mechanical and physical background. In this chapter we utilize PQW to solve the integral equation under certain smoothness conditions on the coefficients and the stiffness matrix being given. We prove that the order of complexity is $O(N)$. This is the lowest order of complexity. The combination of PQW and multi-scale strategy can speed up the computational procedure (cf. [CP2]).

In Chapter IV we construct another kind of periodic wavelets by different method. These wavelets possess a number of nice properties: interpolation, localization, symmetry, regularity up to any order, real value, bi-orthogonality and having explicit representations. Therefore, we believe that they should contain potential value in applications.

The support of the National Sciences Foundation through a grant to me for long period is gratefully acknowledged.

Before the preparation of the manuscript, Dr. Wang Yi-Li and Dr. Li Deng-Feng suggested to me that I should write a book on my research fields. I am grateful to them for their kind suggestion.

I would like to mention Dr. Peng, Si-Long for he not only produced all graphs in this book and corrected typos, but he also made important contribution in Chapter III and IV of this book ([CP2] and [CLPX]), and to him I am very appreciative of his work.

Finally, I am greatly indebted to Ms. Wang Ting for typing most of the manuscript; to Dr. Chen Jing-Yi for modifying the language; to my wife Tian Yu-Xiu for typing the Subject Index and the Author Index, and for xii

supporting me to write the book.

Chen, Han-lin
Professor at the Institute of Mathematics,
Academia Sinica

Chapter 1

THEORY AND APPLICATION OF COMPLEX HARMONIC SPLINE FUNCTIONS

It is well known that complex analysis has many important applications in applied sciences. But for almost all physical systems we cannot write down the explicit expressions of the solution. So we need to construct approximating functions from the given conditions. For instance, the construction of conformal mappings is an important problem both in theoretical study and in practice in various areas. (see p.53, Note 1). In this regard we would like to mention some recent developments in applications of conformal mappings: diffraction of electromagnetic waves, atomic physics, nonlinear diffusion problems, etc.. One can see its applications in various important disciplines (cf. [SL]).

We know that any simply connected domain D contained in a closed Riemann surface, whose boundary γ is a continuum, can be mapped conformally onto the unit disk $|z| < 1$. But if the domain is arbitrarily prescribed it is difficult to obtain the analytic expression of the conformal mapping. For this reason mathematicians have devoted a large amount of work to develop approximations to the mapping functions. We would like to present two different methods to approximate the conformal mapping: the first one uses the so called complex harmonic spline functions which will be introduced in this chapter, and the second is to solve a certain kind of integral equations by applying the quasi-wavelets which will be introduced in Chapter III.

§1.1 The Interpolating Complex Spline Functions on Γ.

Let Γ be the unit circle, $\Delta : z_1, \cdots, z_K$ be points arranged on Γ in counterclockwise order, and $\Phi_n(\Delta)$ denote the family of complex splines of degree n with knots Δ. If $S \in \Phi_n(\Delta)$, then S satisfies the following conditions: (i) $S \in \pi_n$ on γ_j, $j = 1, \cdots, K$, (ii) $S \in C^{n-1}(\Gamma)$, where π_n denotes the family of polynomials with complex variable $z(= e^{i\theta})$ of degree n, γ_j is the circular arc $\gamma(z_j, z_{j+1})$. For clarity we give more explanation of this. Let $I_{a,b}$ be a circular arc with end points $z_a, z_b; z_a \neq z_b$. If a point z runs from z_a to z_b counterclockwise then z describes the arc $I_{a,b}$. If t_1, t_2 belong to $I_{a,b}$, then $t_1 \ll t_2$ means that the point z runs counterclockwise starting from z_a, meets t_1 first, then t_2. In this case we also write $t_2 \gg t_1$. Evidently, we can distinguish the order of any two points t_1, t_2 on $I_{a,b}$, but this is possible only if $z_a \neq z_b$; in other words, $I_{a,b}$ cannot be a contour. For $s, z \in I_{j,j+K}$, we define

$$(s - z)_+^l := \begin{cases} (s - z)^l, & s \gg z \\ 0, & s \ll z \text{ or } s = z \end{cases}$$

where l is any non-negative integer.

In general, we write $I_{j,j+K}$ or $\gamma_{j,j+K}$ instead of $\gamma(z_j, z_{j+K})$.

For any non-negative integer n, we define a circular spline function of degree n as follows:

$$N_{j,n}(z) = (z_{j+n+1} - z_j)[z_j, \cdots, z_{j+n+1}]_s(s - z)_+^n, \quad z \in I_{j,j+n+1} \quad (1.1.1)$$

$j = 0, 1, \cdots, K - 1$. Here $[z_j, \cdots, z_{j+n+1}]f$ is the divided difference of the $(n + 1)$th order of f with respect to points $z_j, \cdots, z_{j+n+1}$ (if $j + n + 1 = K + \nu, \nu \geq 0$, we denote $z_{j+n+1} = z_\nu$). We then call $N_{j,n}$ the complex B-spline, which is a polynomial complex spline of degree n, i.e., $N_{j,n} \in \Phi_n(\Delta)$.

We list some important properties of complex B-splines below (cf. [C1]):

[P1]: $\mathrm{Supp} N_{j,n} = \overline{I_{j,j+n+1}}$, the closure of $I_{j,j+n+1}$,

[P2]: $\{N_{j,n}\}_{j=1}^K$ forms a basis of $\Phi_n(\Delta)$,

[P3]: $N_{j,n}(z) = \dfrac{z - z_j}{z_{j+n} - z_j} N_{j,n-1}(z) + \dfrac{z_{j+n+1} - z}{z_{j+n+1} - z_{j+1}} N_{j+1,n-1}(z)$,
$z \in \Gamma$.

[P4]: $z^l = \sum_{j=1}^k z_j^{(l)} N_{j,n}(z)$, $l = 0, \cdots, n; (z \in \Gamma)$,

$$\binom{n}{l} z_j^{(l)} = \sum \{z_{j+1}^{\alpha_1} \cdots z_{j+n}^{\alpha_n} : \alpha_1 + \cdots + \alpha_n = l, \alpha_i = 0 \text{ or } 1$$
$(i = 1, \cdots, n)\}.$

On occasion we use the B-spline function with equally spaced knots; these are

$$z_j = \omega_l^j, \quad \omega_l = \exp(ih_l), \quad h_l = 2\pi/K(l),$$
$$K(l) = 2^l K, \quad K \geq 2+n, \, l \in \mathbb{N} \qquad (1.1.2)$$

where $\mathbb{N}$ is the set of non-negative integers.

Then, (1.1.1) will be replaced by

$$N_{j,n}^{(l)}(z) = (-1)^{n+1}(\omega_l^{j+n+1} - \omega_l^j)[\omega_l^j, \cdots, \omega_l^{j+n+1}]_s(z-s)_+^n, \qquad (1.1.3)$$

Define the Fourier transform of f by $\mathcal{F}(f)(x) = \int_\Gamma f(z)z^{-x-1}dz$, where $f \in L_2(\Gamma), L_2(\Gamma)$ is the family of square integrable functions on Γ. convolution of two functions f and g in $L_2(\Gamma)$ can be defined as $(f * g)(z) = \int_\Gamma f(z/t)g(t)dt$, then we have

$$\mathcal{F}(f * g)(x) = \mathcal{F}(f)(x) \cdot \mathcal{F}(g)(x-1), \quad x \text{ real number}. \qquad (1.1.4)$$

Define the function $M_{\nu,m}^{(l)}(z)$ as follows

$$M_{0,0}^{(l)}(z) := \begin{cases} 1, & z \in \gamma(1,\omega_l), \\ 0, & \text{otherwise}, \end{cases} \qquad (1.1.5)$$

$$M_{0,m}^{(l)}(z) := M_{0,0}^{(l)} * M_{0,m-1}^{(l)}(z), \qquad (1.1.6)$$

$$M_{\nu,m}^{(l)}(z) := M_{0,m}^{(l)}(\omega_l^{-\nu}z).$$

It can be shown that

$$M_{\nu,m}^{(l)}(z) = M_{\nu,0}^{(l)} * M_{0,m}^{(l)}(z). \qquad (1.1.7)$$

By induction, we have

$$M_{\nu,m}^{(l)}(z) = \frac{(2\pi i)^m}{K_m^{(l)}} N_{\nu,m}^{(l)}(z), \quad z = e^{i\theta} \in \Gamma, \quad m \geq 0. \qquad (1.1.8)$$

The Fourier expansion of (1.1.3) is

$$N_{j,n}^{(l)}(e^{i\theta}) = K_n^{(l)} \sum_{\nu \in \mathbb{Z}} C_\nu^{(l)} C_{\nu-1}^{(l)} \cdots C_{\nu-n}^{(l)} e^{i\nu(\theta-jh_l)}, \qquad (1.1.9)$$

where

$$K_n^{(l)} = \frac{n!(2\pi i)^n}{\prod\limits_{\nu=1}^{n}(\omega_l^\nu - 1)}, \quad C_j^{(l)} = \frac{1 - \omega_l^j}{2\pi ji} \ (j \neq 0), \quad C_0^{(l)} = \frac{1}{K(l)} \text{ and } K_0^{(l)} = 1$$

(refer to [C2] and [Sch1]).

In order to construct the mapping function we shall use complex harmonic spline function, since it provides high accuracy to the approximation. For this purpose, in the first place we have to find the boundary function for the mapping. Here, we offer two different methods. One is the interpolating complex spline on the circle Γ. The other is the complex quasi-interpolating spline function on Γ (see p.54, Note 2).

We first construct the complex interpolating spline of low degree and then the higher order ones. (see p.54, Note 3). In the following we verify the existence of the interpolating complex spline of degree 2 and 3 (cf. [C3]), the knots of each are spaced arbitrarily on Γ.

Transform the unit circle Γ to the whole real axis; the points $\{z_j\}_0^m$ are mapped onto

$$\{x_j\}_0^m, \quad -\infty < x_0 < x_1 < \cdots < x_m < +\infty,$$
$$x_j = \frac{i(1 + z_j)}{1 - z_j} \quad (j = 0, \cdots, m).$$

Without loss of generality we assume that $1 \in$ interior of $\gamma(z_0, z_1)$. We have:

Theorem 1.1.1   Given $\{f_j\}_0^m$ and $n = 2$. If m is even, then there exists one function $S(z)$ in $\Phi_2(\Delta)$, $\Delta = \{z_0, \cdots, z_m\}$, such that

$$S(z_j) = f_j, \quad j = 0, \cdots, m$$

If m is odd then the coefficient matrix $\Delta_2(A)$ is singular.

Before we prove Theorem 1.1.1 we present three equivalent problems.

Since each function $S(z)$ in $\Phi_n(\Delta)$ can be written into the following form

$$S(z) = \begin{cases} P(z), & P(z) \in \pi_n, \quad z \in \gamma_0 \\ P(z) + \sum\limits_{k=1}^{j} C_k(z - z_k)^n, & z \in \gamma_j, \ j = 1, \cdots, m, \end{cases}$$

where $\{C_k\}$ are constants satisfying

$$\sum_{k=1}^{m} C_k(z_0 - z_k)^{n-l} = 0, \quad l = 0, \cdots, n-1, \qquad (1.1.10)$$

and $\gamma_j = \gamma(z_j, z_{j+1})$.

The system of equations

$$\begin{cases} S(z_j) = f_j, & j = 0, \cdots, m \\ \displaystyle\sum_{k=1}^{m} C_k(z_0 - z_k)^{n-l} = 0, & l = 0, \cdots, n-1 \end{cases} \qquad (1.1.11)$$

is equivalent to

$$\begin{cases} t(x_j) = f_j(x_j + i)^n, & j = 0, \cdots, m \\ \displaystyle\sum_{k=1}^{m} \tilde{C}_k(x_0 - x_k)^{n-l} = 0, & l = 0, \cdots, n-1 \end{cases} \qquad (1.1.12)$$

where

$$t(x) = (x + i)^n S\left(\frac{x - i}{x + i}\right),$$

i is the imaginary unit, and

$$\tilde{C}_k = (2i)^n C_k / (i + x_k)^n, \quad k = 0, \cdots, m.$$

The real and imaginary parts of (1.1.12) both are systems of $m + n + 1$ equations, these two systems possess the same matrix, so we need only to consider one of them. We call the solvability of these systems Problem A.

Let $T_n(x_0, \cdots, x_m)$ be a family of real spline functions. If $f(x) \in T_n(x_0, \cdots, x_m)$, then

$$\begin{cases} t(x) = Q(x), & Q(x) \in \pi_n(x), \quad x \in (-\infty, x_0) \text{ and } x \in [x_m, +\infty) \\ t(x) = Q(x) + \displaystyle\sum_{k=0}^{m} C_k(x - x_k)_+^n, & x \in [x_0, x_m) \end{cases} \qquad (1.1.13)$$

and $t(x) \in C^{n-1}(-\infty, +\infty)$, $\pi_n(x)$ is the family of polynomials of degree n with real variable x.

Problem B is the following. Given real data $\{g_j\}_0^m$, does there exist a function $t(x) \in T_n(x_0, \cdots, x_m)$ such that

$$t(x_j) = g_j, \quad j = 0, \cdots, m \tag{1.1.14}$$

Let $B_j^n(x)$, for $j = 0, \cdots, m-n-1$, denote the B-spline of degree n with support (x_j, x_{j+n+1}) (see [Mars]). Let $B = \text{span}\{x^j \ (j = 0, \cdots, n), \ B_k^n(x) \ (k = 0, \cdots, m - n - 1)\}$. The problem of whether the equations

$$u(x_j) = 0, \quad j = 0, \cdots, m, \quad u(x) \in B \tag{1.1.15}$$

have only the solution $\{0\}$ is called Problem C.

The coefficient matrices in Problems A, B and C will be denoted by $\triangle_n(A)$, $\triangle_n(B)$ and $\triangle_n(C)$ respectively. By simple calculations, we conclude that $\triangle_n(A)$ is non-singular iff $\triangle_n(B)$ is non-singular and the later is equivalent to $\triangle_n(C)$ being non-singular.

The proof of Theorem 1.1.1.

Write $B_{j,k} := B_j^2(x_k)$, then for $n = 2$, the determinant of Problem C is

$$|\triangle_2(C)| = \begin{vmatrix} B_{0,1} & 0 & \cdots & 0 & (x_0 - x_1) & (x_0 - x_1)^2 \\ B_{0,2} & B_{1,2} & 0 & \cdots & (x_0 - x_2) & (x_0 - x_2)^2 \\ \cdots & \cdots & \cdots & \cdots & \cdots & \cdots \\ \cdots & \cdots & \cdots & \cdots & \cdots & \cdots \\ 0 & \cdots & 0 & B_{m-3,m-1} & (x_0 - x_{m-1}) & (x_0 - x_{m-1})^2 \\ 0 & \cdots & & 0 & (x_0 - x_m) & (x_0 - x_m)^2 \end{vmatrix}$$

$$= \begin{cases} 0, & m \quad \text{odd} \\ -(x_m - x_0)^2 (x_2 - x_1) \displaystyle\prod_{j=0}^{m-3} B_{j,j+2} < 0, & m \quad \text{even} \end{cases}$$

This completes the proof of Theorem 1.1.1, □.

Theorem 1.1.2 Let $\{z_j\}_0^m$ be $m + 1$ points spaced arbitrarily on Γ. Given $m + 1$ data $\{f_j\}_0^m$, then there is a unique function $S(z)$ in $\Phi_3(\triangle)$,

$\Delta = \{z_0, \cdots, z_m\}$, such that

$$S(z_j) = f_j \qquad j = 0, \cdots, m.$$

Proof.   We need to investigate Problem B. Let $m_j = t^{(2)}(x_j), j = 0, \cdots, m$. Then we have the following equations:

$$\begin{cases} \dfrac{h_j}{6} m_{j-1} + \dfrac{h_j + h_{j+1}}{3} m_j + \dfrac{h_{j+1}}{6} m_{j+1} \\[2mm] \qquad = \dfrac{y_{j+1} - y_j}{h_{j+1}} - \dfrac{y_j - y_{j-1}}{h_j} \\[2mm] \qquad j = 1, \cdots, m-1; \quad h_j = x_j - x_{j-1}, \\[2mm] 2(x_m - x_1)m_0 - (x_1 - x_0)m_1 + (x_m - x_0)m_m \\[2mm] \qquad = 6 \left(\dfrac{y_m - y_0}{x_m - x_0} - \dfrac{y_1 - y_0}{x_1 - x_0} \right), \\[2mm] (x_m - x_0)m_0 - (x_m - x_{m-1})m_{m-1} + 2(x_{m-1} - x_0)m_m \\[2mm] \qquad = 6 \left(\dfrac{y_m - y_{m-1}}{x_m - x_{m-1}} - \dfrac{y_m - y_0}{x_m - x_0} \right), \end{cases} \qquad (1.1.16)$$

$\{y_j\}_0^m$ being the given data.

The above equations are obtained from the following system of equations:

$$t(x_j) = y_j, \quad j = 0, \cdots, m,$$
$$\sum_{k=1}^{m} C_k(x_0 - x_k)^l = 0, \quad l = 1, 2, 3. \qquad (1.1.17)$$

where $t(x) \in T_3(x_0, \cdots, x_m)$.

Denote $\Delta(x_0, \cdots, x_m)$ the coefficient determinant of (1.1.16), then

$$\Delta(x_0, \cdots, x_m) = \left(\sum_{j=1}^{m-1} h_j \right) \prod_{k=1}^{m-2} (h_k + h_{k+1}) \tilde{\Delta},$$

$\tilde{\triangle} = (h_{m-1} + h_m)(h_2 + \cdots + h_m)\triangle^*$, and

$$\triangle^* = \begin{vmatrix} 2 & -\lambda_1 & & & & & & 1+\lambda_1 \\ 1-\lambda_2 & 2 & \lambda_2 & & & & & \\ & 1-\lambda_3 & 2 & \lambda_3 & & & & \\ & & & \ddots & & & & \\ & & & & \ddots & & & \\ & & & & & \ddots & & \\ & & & & 1-\lambda_{m-1} & 2 & \lambda_{m-1} & \\ & & & & & 1-\lambda_m & 2 & \lambda_m \\ 1+\lambda_{m+1} & & & & & & -\lambda_{m+1} & 2 \end{vmatrix}$$

where $\lambda_1 = h_1/(h_2 + \cdots + h_m), \lambda_{m+1} = h_m/(h_1 + \cdots + h_{m-1}), \lambda_j = h_j/(h_{j-1}+h_j), j = 2, \cdots, m$. It can be shown that $\frac{\partial \tilde{\triangle}}{\partial h_m} > 0$, $\lim\limits_{x_m \to x_{m-1}} \triangle(x_0, \cdots, x_m) > 0$ and $\lim\limits_{x_m \to \infty} \triangle(x_0, \cdots, x_m) > 0$. Therefore, we have $\triangle(x_0, \cdots, x_m) > 0$. $\square$.

Remark (i) The detailed proof of Theorem 1.1.1 and Theorem 1.1.2 can be found in [C3]. (ii) In fact, the proof of the existence and uniqueness involves the construction of the interpolating spline function.

For the interpolation by high order complex spline functions with some deficiency, we have to generalize the definition.

Take a non-decreasing sequence $t = \{t_j\}_{-\infty}^{\infty}$ with

$$t_j < t_{j+n+1}, \quad t_{j+K} = t_j + 2\pi \quad (j \in \mathbb{Z}) \tag{1.1.18}$$

and write $z_j = e^{it_j} (j \in \mathbb{Z})$. We now define $\Phi_n(\underset{\sim}{t}, \mu)$ to be the class of all functions on Γ which coincide with an element of π_n on each non-trivial arc $\gamma(z_j, z_{j+1})$ and which are $C^{n-\mu}$ at z_j if t_j has multiplicity μ in $\underset{\sim}{t}$.

It is clear that $\Phi_n(\underset{\sim}{t}, 1) = \Phi_n(\triangle)$.

Take a non-decreasing sequence $\underset{\sim}{\tau} = \{t_j\}_{-\infty}^{\infty}$ which satisfies

$$\tau_j < \tau_{j+n+1}, \quad \tau_{j+K} = \tau_j + 2\pi \quad (j \in \mathbb{Z}) \tag{1.1.19}$$

and let $w_j = e^{i\tau_j} (j \in \mathbb{Z})$. We say that $(\Phi_n(\underset{\sim}{t}, \mu), \underset{\sim}{\tau})$ is solvable if for every

sequence $\underset{\sim}{y} = (y_1, \cdots, y_K)$, there is a unique function S in $\Phi_n(\underset{\sim}{t}, \mu)$ which

interpolates $\underset{\sim}{y}$ at $(w_1, \cdots, w_K)$, i.e.,

$$S^{(\nu_j)}(w_j) = y_j \quad (j = 1, \cdots, K) \qquad (1.1.20)$$

where

$$\nu_j = |\{k < j : \tau_k = \tau_j\}|. \qquad (1.1.21)$$

Theorem 1.1.3 If $(\Phi_n(\underset{\sim}{t}, \mu), \underset{\sim}{\tau})$ is solvable, then there exists $l \in \mathbb{Z}$
with

$$t_i < \tau_{i+l} < t_{i+n+l}, \quad (i \in \mathbb{Z}). \qquad (1.1.22)$$

Moreover,

(A) if $n + K$ is odd and (1.1.22) holds, then $(\Phi_n(\underset{\sim}{t}, \mu), \underset{\sim}{\tau})$ is solvable;

(B) if $n + K$ is even, then

    a) if $\underset{\sim}{\tau}$ is strictly increasing and $\tau_j = t_j (j \in \mathbb{Z})$,   (1.1.23)

    then $(\Phi_n(\underset{\sim}{t}, \mu), \underset{\sim}{\tau})$ is solvable if and only if n is odd;

    b) suppose

$$t_j < \tau_j < t_{j+1} \quad (j \in \mathbb{Z}) \qquad (1.1.24)$$

is satisfied, and for some $l, \tau_{j+l} = \tau_j + \pi(j \in \mathbb{Z})$. Then $(\Phi_n(\underset{\sim}{t}, \mu), \underset{\sim}{\tau})$ is
solvable if and only if n is even;

    c) suppose (1.1.24) is satisfied and t and τ are symmetric about the
same point. Then $(\Phi_n(\underset{\sim}{t}, \mu), \underset{\sim}{t})$ is solvable if and only if n is even.

The proof of Theorem 1.1.3 (See [GL1], [GL2]) is quite different from
that of Theorem 1.1.1 and Theorem 1.1.2. As for the uniform knots and
interpolatory points, we have the following:

Corollary 1.1.1 If $\underset{\sim}{t}$ and $\underset{\sim}{\tau}$ are uniform, i.e., $t_{j+1} - t_j = \tau_{j+1} -$
$\tau_j = 2\pi/K (j \in \mathbb{Z})$, then $(\Phi_n(\underset{\sim}{t}, \mu), \underset{\sim}{\tau})$ is solvable except in the following
cases:

    a) n and K are even, and for some $\nu, \tau_{j+\nu} = t_j$,

b) n and K are odd, and for some $\nu, \tau_{j+\nu} = \frac{1}{2}(t_j + t_{j+1}), j \in \mathbb{Z}$, (cf. [TZ]).

Now we would like to ask: how fast does the interpolating function approximate a given function?

In the following we shall give the estimates for the errors between the approximated functions and the interpolating cubic spline functions. Note that in practical applications, the cubic spline function is often used.

Theorem 1.1.4   Let $f(z)$ be continuous on Γ, and $s(z)$ be the complex cubic spline with knots $\{z_j\}_1^N$ such that $s(z_j) = f(z_j)(j = 1, \cdots, N)$. Then we have

$$|s(z) - f(z)| < K(R)\omega(f, |\triangle|), \quad z \in \Gamma. \tag{1.1.25}$$

where $R = \max_j |z_{j+1} - z_j| / \min_j |z_{j+1} - z_j|, \omega(f, |\triangle|)$ is the modulus of continuity of f on Γ, $K(R) = \min(5.13R + 7.13, 0.07R^2 + 1.5)$ and $|\triangle| = \max_j |z_{j+1} - z_j|$ (The proof can be found in [CH1]).

Proof.   For z on γ_{j-1}, we have

$$\begin{aligned}
S(z) = &\frac{m_{j-1}}{h_{j-1}^2}(z_j - z)^2(z - z_{j-1}) - \frac{m_j}{h_{j-1}^2}(z - z_{j-1})^2(z_j - z) \\
&+ \frac{f_{j-1}}{h_{j-1}^3}(z_j - z)^2[2(z - z_{j-1}) \\
&+ h_{j-1}] + \frac{f_j}{h_{j-1}^3}(z - z_{j-1})^2[2(z_j - z) + h_{j-1}]
\end{aligned} \tag{1.1.26}$$

where $h_{j-1} = z_j - z_{j-1}, f_j = S(z_j), m_j = S'(z_j)$.

The quantities m_j satisfy the requirement that $S^{(2)}(z_{j-}) = S^{(2)}(z_{j+})$ for $j = 1, \cdots, N$:

$$a_j m_{j-1} + 2m_j + b_j m_{j+1} = Q_j \tag{1.1.27}$$

where

$$a_j = \frac{h_j}{h_{j-1} + h_j}, \quad b_j = 1 - a_j, \quad Q_j = 3a_j \frac{f_j - f_{j-1}}{h_{j-1}} + 3b_j \frac{f_{j+1} - f_j}{h_j}$$

(1.1.27) may be written in matrix form as

$$Am = Q. \tag{1.1.28}$$

Since we can prove the following inequalities:

$$\|A^{-1}\| \leq \frac{1}{\min_j(2 - |a_j| - |b_j|)} \leq \frac{1}{2 - \sqrt{2}}, \tag{1.1.29}$$

$$\lambda_j(z) = \frac{|(z - z_{j-1})(z - z_j)^2|}{|h_{j-1}|^2} \leq \frac{|h_{j-1}|}{2} \qquad \text{for} \quad z \in \gamma_{j-1},$$

$$\mu_j(z) = \frac{|(z - z_{j-1})^2(z - z_j)|}{|h_{j-1}|^2} \leq \frac{|h_{j-1}|}{2} \qquad \text{for} \quad z \in \gamma_{j-1},$$

$$g(\lambda) = \left| \frac{3\lambda}{2h_{j-1}} - \frac{2\lambda^3}{h_{j-1}^3} \right| \leq 1, \quad \lambda = z - \tfrac{1}{2}(z_{j-1} + z_j), \quad z \in \gamma_{j-1},$$

it follows from (1.1.28) and (1.1.29) that

$$|m_j| \leq \max_k |Q_k| \sum_{\nu=1}^{N} |A_{j\nu}^{-1}| \leq \frac{\max_k |Q_k|}{2 - \sqrt{2}}. \tag{1.1.30}$$

For $z \in \gamma_{k-1}$, in view of (1.1.26), (1.1.29) and (1.1.30), we have

$$|S(z) - f(z)| \leq \max_k |m_k|(\lambda_k(z) + \mu_k(z)) + g(\lambda))|f_k - f_{k-1}|$$

$$+ \left| \frac{f_k + f_{k-1}}{2} - f(z) \right|$$

$$\leq \frac{\max_k |Q_k|}{2 - \sqrt{2}} |h_{k-1}| + \left| \frac{f_k + f_{k-1}}{2} - f(z) \right| + |f_k - f_{k-1}|. \tag{1.1.31}$$

From (1.1.31) we can easily prove the following inequality

$$|S(z) - f(z)| \leq (5.13R + 7.13)\omega(f, |\triangle|). \tag{1.1.32}$$

On the other hand, the continuity of $S'(z)$ at z yields the following identity

$$b_j M_{j-1} + 2M_j + a_j M_{j+1} = P_j, \quad j = 1, \cdots, N \tag{1.1.33}$$

where

$$M_j = S^{(2)}(z_j), \quad P_j = \frac{1}{6(h_{j-1} + h_j)} \left[\frac{f_{j+1} - f_j}{h_j} - \frac{f_j - f_{j-1}}{h_{j-1}} \right].$$

(1.1.33) can be written in matrix form as

$$BM = P, \tag{1.1.34}$$

Since we can prove the following inequalities:

$$\|B^{-1}\| \le (2 - \sqrt{2})^{-1} \tag{1.1.35}$$

$$|\alpha_j| = |(z_j - z)[(z_j - z)^2 - h_{j-1}^2]|\,|6h_{j-1}|^{-1}$$

$$\le \frac{|h_{j-1}|^2}{12}, \quad \text{for } z \in \gamma_{j-1},$$

$$|\beta_j| = |(z - z_{j-1})[(z - z_{j-1})^2 - h_{j-1}^2]|\,|6h_{j-1}|^{-1}$$

$$\le \frac{|h_{j-1}|^2}{12}, \quad \text{for } z \in \gamma_{j-1},$$

$$|P_j| \le \frac{\omega(f, |\triangle|)}{3\sqrt{2}|h_j h_{j-1}|},$$

it follows from (1.1.34) and (1.1.35) that

$$|M_j| \le \frac{(1 + \sqrt{2})\omega(f, |\triangle|)}{6 \min_k |h_k h_{k-1}|}, \quad j = 1, \cdots, N. \tag{1.1.36}$$

In view of (1.1.35) and (1.1.36), for $z \in \gamma_j$, we have

$$
\begin{aligned}
|S(z) - f(z)| &= \left| \alpha_j M_{j-1} + \beta_j M_j + \left(\frac{f_j + f_{j-1}}{2} - f(z) \right) \right. \\
&\quad \left. -(f_j - f_{j-1}) \cdot \frac{z_j + z_{j-1} - 2z}{2h_{j-1}} \right| \\
&\le \left(\frac{3}{2} + \frac{1 + \sqrt{2}}{36} R^2 \right) \omega(f, |\triangle|) \\
&\le (1.5 + 0.07 R^2)\omega(f, |\triangle|).
\end{aligned}
\tag{1.1.37}
$$

(1.1.25) now follows from (1.1.36) and (1.1.37).   □.

A closed Jordan curve γ is said to satisfy the Ljapunov condition if it can be represented by $\zeta = \zeta(s), 0 \le s < L$, where L is the length of γ and s is the arc length of γ, measured from a fixed point, such that if s varies from 0 to L, then $\zeta(s)$ makes one turn on γ in the positive sense, γ has a tangent at every point, which varies continuously, and $\zeta'(s)$ satisfies the following Hölder condition:

$$|\zeta'(s_1) - \zeta'(s_2)| \le J|s_1 - s_2|^\alpha, \quad 0 < \alpha \le 1, \quad J = \text{const.}$$

If the function $w = F(z)$ maps the unit disk U conformally onto D and $\gamma = \partial D$ satisfies the Ljapunov condition, then $F'(z)$ exists in $\overline{U}$, is different from zero, and satisfies the same Hölder condition:

$$|F'(e^{i\theta_1}) - F'(e^{i\theta_2})| \leq K|\theta_1 - \theta_2|^\alpha, \quad K = \text{const}$$

(see [LS], [T]).

Theorem 1.1.5 Assume γ satisfies the Ljapunov condition and $w = F(z)$ maps U conformally onto $D, \partial D = \gamma$. If $S(z)$ is the complex cubic spline which interpolates $F(z)$ at $\triangle = \{z_j\}_1^N$—the knots of $S(z)$—then, for $z \in \Gamma, \Gamma = \partial U$, we have

$$|S(z) - F(z)| \leq K_1| \triangle |^{1+\alpha}, \quad z \in \Gamma, \tag{1.1.38}$$

$$|S'(z) - F'(z)| \leq K_2| \triangle |^\alpha, \quad z \in \Gamma, \tag{1.1.39}$$

$$\left| \frac{(S'(t_1) - F'(t_1)) - (S'(t_2) - F'(t_2))}{| \triangle |^{\alpha-\delta}} \right| \leq K_3|t_1 - t_2|^\delta, \quad t_1, t_2 \in \Gamma \tag{1.1.40}$$

for any $\delta, 0 < \delta < \alpha$, where R is the mesh ratio and

$$| \triangle | = \max_j |z_j - z_{j-1}|, \quad K_1 = \frac{\pi}{2}K_2, \quad K_2 = (26 + 14\sqrt{2})\frac{K}{\alpha+1}\left(\frac{\pi}{2}\right)^{\alpha+1},$$

$$K_3 = 2\left(\frac{38 + 40\sqrt{2}}{\alpha+1} \cdot \frac{\pi}{2} + 1\right)\left(\frac{\pi}{2}\right)^\alpha KR.$$

(see [CH1]).

Proof.   From (1.1.28)

$$A\left(m - \frac{1}{3}Q\right) = (3I - A)\frac{1}{3}Q. \tag{1.1.41}$$

The jth element of the vector $(3I - A)\frac{1}{3}Q$ may be estimated as follows. Since

$$\left| \frac{1}{h_j} \int_{z_j}^{z_{j+1}} (F'(t) - F'(z_{j+1}))dt \right| \leq \frac{K}{\alpha+1}\left(\frac{\pi}{2}\right)^{\alpha+1}| \triangle |^\alpha \tag{1.1.42}$$

we have

$$[(3I - A) - Q]_j = \left| \frac{-a_j Q_{j-1}}{3} + \frac{Q_j}{3} - \frac{b_j Q_{j+1}}{3} \right|$$

$$\leq 4 \max_k \left| \frac{1}{h_k} \int_{z_k}^{z_{k+1}} (F'(t) - F'(z_{k+1})) dt \right| \leq \frac{4K}{\alpha+1} \left(\frac{\pi}{2} \right)^{\alpha+1} |\Delta|^\alpha.$$

(1.1.43)

From (1.1.41), we have

$$\left| m_j - \left(b_j \frac{F_{j+1} - F_j}{h_j} + a_j \frac{F_j - F_{j-1}}{h_{j-1}} \right) \right|$$

$$\leq \|A^{-1}\| \max_k \left[(3I - A) \frac{1}{3} Q \right]_k \left(\frac{4K(\pi/2)^{\alpha+1}|\Delta|^\alpha}{(\alpha+1)(2-\sqrt{2})} \right).$$

(1.1.44)

Therefore, from (1.1.42), (1.1.44) and $a_j + b_j = 1$, we find

$$\left| m_j - \frac{F_j - F_{j-1}}{h_{j-1}} \right|$$

$$\leq \left| m_j - \left(b_j \frac{F_{j+1} - F_j}{h_j} + a_j \frac{F_j - F_{j-1}}{h_{j-1}} \right) \right|$$

$$+ \left| \frac{1}{h_j} \int_{z_j}^{z_{j+1}} (F'(t) - F'(z_j)) dt - \frac{1}{h_{j-1}} \int_{z_{j-1}}^{z_j} (F'(t) - F'(z_j)) dt \right|$$

$$\leq \left(\frac{2}{2-\sqrt{2}} + 1 \right) \frac{2K}{\alpha+1} \left(\frac{\pi}{2} \right)^{\alpha+1} |\Delta|^\alpha.$$

(1.1.45)

Furthermore,

$$\left| S'(z) - \frac{F_j - F_{j-1}}{h_{j-1}} \right| \leq \left| \frac{3\varepsilon^2}{h_{j-1}^2} - \frac{1}{4} \right| \left| m_{j-1} + m_j - 2 \frac{F_j - F_{j-1}}{h_{j-1}} \right|$$

$$+ \left| \frac{\varepsilon}{h_{j-1}} \right| |m_j - m_{j-1}|, \quad \text{for} \quad z \in \gamma_{j-1},$$

(1.1.46)

where $\varepsilon = \frac{z_j + z_{j-1}}{2} - z_j = \tilde{z} - z_j$. Evidently,

$$\varepsilon \leq \frac{|h_{j-1}|}{\sqrt{2}}, \quad \left| \frac{3\varepsilon^2}{h_{j-1}^2} - \frac{1}{4} \right| \leq \frac{7}{4}.$$

(1.1.47)

Let

$$Y_j = \frac{1}{h_j} \int_{z_j}^{z_{j+1}} (F'(t) - F'(z_{j+1})) dt,$$

then, from (1.1.42) and (1.1.44),

$$|m_j - m_{j-1}|$$

$$\leq \left| m_j - \frac{F_j - F_{j-1}}{h_{j-1}} - \left(m_{j-1} - \frac{F_{j-1} - F_{j-2}}{h_{j-2}} \right) \right|$$

$$+ |Y_{j-1} - Y_{j-2}| \tag{1.1.48}$$

$$\leq \left(\frac{4}{2 - \sqrt{2}} + 3 \right) \times \frac{2K}{\alpha + 1} \left(\frac{\pi}{2} \right)^{\alpha+1} |\triangle|^{\alpha}.$$

Combining (1.1.46)–(1.1.48), we obtain

$$\left| S'(z) - \frac{F_j - F_{j-1}}{h_{j-1}} \right| \leq (25 + 14\sqrt{2}) \frac{K}{\alpha + 1} \left(\frac{\pi}{2} \right)^{\alpha+1} |\triangle|^{\alpha}, \quad z \in \gamma_{j-1},$$

$$\tag{1.1.49}$$

Thus, for $z \in \gamma_{j-1}$,

$$|S'(z) - F'(z)| \leq \left| S'(z) - \frac{F_j - F_{j-1}}{h_{j-1}} \right| + \left| \frac{1}{h_{j-1}} \int_{z_{j-1}}^{z_j} (F'(z) - F'(t)) dt \right|$$

$$\leq (26 + 14\sqrt{2}) \frac{K}{\alpha + 1} \left(\frac{\pi}{2} \right)^{\alpha+1} |\triangle|^{\alpha}.$$

Inequality (1.1.39) is proved.

Inequality (1.1.38) follows from (1.1.39) by integration.

To prove (1.1.40), we notice that

$$|S'(z) - S'(t)|$$

$$= |z - t| \left| \left[\frac{m_{j-1} - \dfrac{F_j - F_{j-1}}{h_{j-1}}}{h_{j-1}} + \frac{m_j - \dfrac{F_j - F_{j-1}}{h_{j-1}}}{h_{j-1}} \right] \right. \tag{1.1.50}$$

$$\left. \times \left(\frac{3(z + t) - 6\tilde{z}}{h_{j-1}} \right) + \frac{m_j - m_{j-1}}{h_{j-1}} \right|,$$

$$\tilde{z} = \frac{1}{2}(z_{j-1} + z_j), \quad t, z \in \gamma_{j-1},$$

from (1.1.45), (1.1.48), (1.1.50) and the estimate $|(3(z + t) - 6\tilde{z})/h_{j-1}| \leq 3\sqrt{2}$, we have

$$|S'(z) - S'(t)| \leq \frac{2(19 + 20\sqrt{2})}{\alpha + 1} \left(\frac{\pi}{2} \right)^{\alpha+1} K |\triangle|^{\alpha} \left| \frac{z - t}{h_{j-1}} \right|.$$

Therefore, if $z, t \in \gamma_{j-1}$, we have

$$\frac{|(S'(z) - F'(z)) - (S'(t) - F'(t))|}{|\triangle|^{\alpha-\delta}} \leq \left[\frac{19 + 20\sqrt{2}}{\alpha+1}\pi R + 1\right]\left(\frac{\pi}{2}\right)^\alpha K|z - t|^\delta,$$

$$(1.1.51)$$

where $R = |\triangle|/\min_j|z_j - z_{j-1}|$.

If, on the other hand, $z \in \gamma_j, t \in \gamma_\nu, j \neq \nu$, we distinguish three cases:

(1) If $|t - z| \geq |\triangle|$, then from (1.1.39),

$$\frac{|(S'(z) - F'(z)) - (S'(t) - F'(t))|}{|\triangle|^{\alpha-\delta}} \leq 2K_2|\triangle|^\delta \leq K_3|t - z|^\delta. \quad (1.1.52)$$

(2) If $|t - z| < |\triangle|, |z - t| \geq \min_j|h_j|$, then from (1.1.39),

$$\frac{|(S'(z) - F'(z)) - (S'(t) - F'(t))|}{|\triangle|^{\alpha-\delta}} \leq 2K_2|\triangle|^\delta \leq K_3|t - z|^\delta. \quad (1.1.53)$$

(3) If $|t - z| < |\triangle|$ and $|t - z| < \min|h_j|$, we assume, without loss of generality, that $t \in \gamma_{j+1}, z \in \gamma_j$. Then

$$\frac{|(S'(z) - F'(z)) - (S'(t) - F'(t))|}{|\triangle|^{\alpha-\delta}}$$

$$\leq \frac{|(S'(z) - F'(z)) - (S'(z_{j+1}) - F'(z_{j+1}))|}{|\triangle|^{\alpha-\delta}},$$

$$\frac{|(S'(t) - F'(t)) - (S'(z_{j+1}) - F'(z_{j+1}))|}{|\triangle|^{\alpha-\delta}}$$

$$\leq K'(|z - z_{j+1}|^\delta + |t - z_{j+1}|^\delta)$$

$$\leq 2K'|z - t|^\delta \leq K_3|z - t|^\delta \quad (1.1.54)$$

where $K' = \left[\dfrac{19 + 20\sqrt{2}}{\alpha+1}\pi R + 1\right]\left(\dfrac{\pi}{2}\right)^\alpha K$, the second inequality being obtained from (1.1.51).

From (1.1.51)–(1.1.54) formula (1.1.40) follows.   □.

In section 1.1 we have pointed out the existence of the interpolating complex spline functions, but the construction of each of them is based on the solution of a system of linear equations. For low degree splines (i.e., $n = 2$ and $n = 3$), the proof of the existence involves the construction of interpolating splines as pointed before, but for the case of splines of higher degree, it is rather elaborate to calculate the coefficients for the solution.

In the following we shall present a totally different but rather simple method to construct the quasi-interpolant splines, which provide an efficient approximation to the given function.

§1.2   Quasi-interpolant Complex Splines on Γ.

Define an operator $\mathcal{L}$ as follows. Its domain is $C^{n-1}(\Gamma)$ and it satisfies the following two conditions:

$$\mathcal{L}(g) = \sum_{j=1}^{K}(L_j g)N_{j,n} \quad \forall g \in C^{n-1}(\Gamma), \tag{1.2.1}$$

where $\{N_{j,n}\}_1^K$ are defined in the beginning of §1.1;

$$L_j(g) = \sum_{r \leq n} T_{j,r}g^{(r)}(t_j), \quad j = 1, \cdots, K, \quad t_j \in I_{j,j+n+1} \tag{1.2.2}$$

where $T_{j,r}$ are constants, and $g^{(r)}$ is the rth derivative of g with respect to z.

Follow the idea described in the real case [BF] we have the following important theorem about the operator $\mathcal{L}$.

Theorem 1.2.1   Let $\mathcal{L}$ be the operator defined by (1.2.1) and (1.2.2). If anyone of the following three proposition (A), (B), (C) is valid, then the other two are true.

(A) $\mathcal{L}(S) = S$, for any $S \in \Phi_n(\Delta)$.

(B) $L_j(N_{i,n}) = \delta_{j,i}, \ i,j = 1, \cdots, K$ ($\delta_{i,j}$, the Kronecker delta).

(C) $T_{j,r} = ((-1)^{n-r}/n!)\lambda_j^{(n-r)}(t_j), r = 0, \cdots, n, \ j = 1, \cdots, K,$ with

$$\lambda_j(z) = \prod_{\nu=j+1}^{j+n} (z_\nu - z), \ t_j \in I_{j,j+n+1}.$$

Hereafter we stipulate that the operator $\mathcal{L}$ of the form (1.2.1) satisfies one of the three conditions (A), (B) and (C), hence all.

This Theorem 1.2.1 is extracted from [C4].

Corollary $\mathcal{L}(P) = P$, for all $P \in \pi_n :=$ the family of polynomials of degree n.

We now begin to study the error of the quasi-interpolation procedure. Before doing this we have to establish two lemmas.

Lemma 1.2.1   Assume that the length of each interval $I_{i,i+n+1}$ is less than $\pi : |I_{i,i+n+1}| < \pi (i = 1, \cdots, K)$, and $Z_{i+K} = Z_i (i = 1, \cdots, K)$. Then

(a) $N_{s,n}(z) \neq 0, z \in I_{s,s+n+1}$

(b) $|N_{s,n}(z)| \leq 2^n, z \in \Gamma$

Proof. Let $z' = ze^{i\theta}, z'_j = z_j e^{i\theta}, (j = 1, \cdots, K)$, where θ is a real number $0 \leq \theta \leq \pi$. If $\tilde{N}_{s,n}(z')$ denotes a complex B-spline of degree n with knots $\{z'_j\}_s^{s+n+1}$, we have $N_{s,n}(z) = \tilde{N}_{s,n}(z')$. Thus the values of a complex spline do not change under rotations. Therefore we may assume that the point $z = 1$ does not belong to the support of $N_{s,n}(z)$, i.e. $1 \in \Gamma \backslash I_{s,s+n+1}$.

Under the transformation $z = (x - i)/(x + i)$, we obtain

$$N_{s,n}(z) = B_s^n(x) \prod_{\nu=s+1}^{s+n} \frac{(x_\nu + i)}{(x + i)^n}, \qquad (1.2.3)$$

where $B_s^n(x)$ is a real B-spline of degree n with real knots $\{x_j\}_{j=s}^{s+n+1}, x_j = i((1 + z_j)/(1 - z_j))$. From (1.2.3) we obtain (a).

By §1.1, [P3], we have (b).   □.

From Lemma 1.2.1, (b) and the definition of the complex B-spline we have the following:

Lemma 1.2.2   For $z \in I_{i,i+n+1}$, we have

$$|N_{i,n}^{(r)}(z)| \leq \frac{2^n n!}{(n - r)! \eta_{i+n-1} \eta_{i+n-2} \cdots \eta_{i+n-r}}, \qquad r = 1, \cdots, n$$

where $\eta_{i+n-\mu} = \min_{0 \leq \nu \leq \mu} |z_{i+\nu+n+1-\mu} - z_{i+\nu}|, 1 \leq \mu \leq r, r \leq n$.

Let

$$E = \mathcal{L}(f) - f,$$

$$Y_z f = \sum_{r=0}^{n} f^{(r)}(z)(\cdot - z)^r / r! \in \pi_n,$$

$$f = Y_z f + R_z.$$

Evidently,

$$E^{(s)}(z) = \frac{d^s(\mathcal{L}(R_z))}{dz^s}, \quad 0 \le s \le n. \tag{1.2.4}$$

The integral representation for the remainder $R_z(t)$ has the following form:

$$R_z^{(r)}(t) = \frac{1}{(n-1-r)!} \int_z^t R_z^{(n)}(\eta)(t-\eta)^{n-1-r} d\eta \quad (0 \le r \le n) \tag{1.2.5}$$

$f^{(n)}$ is absolutely continuous on Γ,

or

$$R_z^{(r)}(t) = \frac{1}{(n-r)!} \int_z^t R_z^{(n+1)}(\eta)(t-\eta)^{n-r} d\eta \quad (0 \le r \le n) \tag{1.2.6}$$

$f^{(n+1)}$ is continuous on Γ.

If the arc length $| \widehat{zt} |$ is less than π, we have:

$$| \widehat{zt} | < \frac{\pi}{2} |z - t|. \tag{1.2.7}$$

We suppose all the arc lengths $|\bar{I}_{i,i+n+1}|$ are less than π, $i = 1, \cdots, K$. From Lemma 1.2.1 and 1.2.2, and (1.2.4)–(1.2.7) we have

Theorem 1.2.2   Let $f^{(n)}$ be absolutely continuous on Γ. Let $\mathcal{L}$ be the quasi-interpolant operator defined by (1.2.1) and (1.2.2) satisfying one of the three conditions (A), (B) and (C). Let $t_j \in \bar{I}_{j+\lambda,j+n+1-\lambda}$, for $\lambda = \lfloor (n+1)/2 \rfloor$. If $E := \mathcal{L}(f) - f$, then

$$|E^{(s)}(z)| \le K_s \omega(f^{(n)}; |\triangle|)|\triangle|^{n-s}, \quad 0 \le s \le n \tag{1.2.8}$$

where

$$|\triangle| = \max_{1 \le j \le K} |z_{j+1} - z_j|, \quad \omega(g,h) = \sup_{\substack{|t_1 - t_2| \le h \\ t_1, t_2 \in \Gamma}} |g(t_1) - g(t_2)|.$$

If $s \leq \lfloor (n+1)/2 \rfloor$, then K_s is a constant independent of the mesh ratio defined by (1.2.11).
(see [C1]).

Theorem 1.2.3   If $f^{(n)}$ satisfies a Lipschitz condition of order α ($0 < \alpha \leq 1$), i.e., $|f^{(n)}(z_1) - f^{(n)}(z_2)| < D|z_1 - z_2|^\alpha$, then

$$|E^{(s)}(z)| < J_s|\triangle|^{n+\alpha-s}, \quad 0 \leq s \leq n \qquad (1.2.9)$$

If $s \leq \lfloor (n+1)/2 \rfloor$, J_s is a constant independent of mesh ratio.

Theorem 1.2.4   If $f^{(n+1)}$ is continuous on Γ, then

$$|E^{(s)}(z)| \leq P_s\|f^{(n+1)}\|_\infty|\triangle|^{n+1-s}, \quad 0 \leq s \leq n. \qquad (1.2.10)$$

If $s \leq \lfloor (n+1)/2 \rfloor$, P_s is a constant independent of mesh ratio.

We now estimate the constants K_s, J_s and P_s in (1.2.8), (1.2.9) and (1.2.10) respectively.

Let

$$\beta := \frac{\max_{1 \leq j \leq K} |z_{j+1} - z_j|}{\min_{1 \leq j \leq K} |z_{j+1} - z_j|} \qquad (1.2.11)$$

From (1.2.4), (1.2.5), (1.2.6), (1.2.7) and Lemma 1.2.2, direct calculation leads to

Theorem 1.2.5   The numbers K_s, J_s, P_s in (1.2.8), (1.2.9), (1.2.10) can be estimated as follows. For $s \leq \lambda$,

$$K_s < \frac{\pi}{2}\left[\frac{\pi}{2}\left(\frac{n+2}{2}\right) + 1\right]C_{s,1} \qquad (1.2.12)$$

$$J_s < \frac{\pi}{2}\left(\frac{n+2}{2}\right)^\alpha C_{s,1} \qquad (1.2.13)$$

$$P_s < \frac{\pi}{4n}(n+2)C_{s,1} \qquad (1.2.14)$$

while for $s > \lambda$, we have

$$K_s < \frac{\pi}{2}\left[\frac{\pi}{2}\left(\frac{n+2}{2}\right)+1\right]C_{s,2} \tag{1.2.15}$$

$$J_s < \frac{\pi}{2}\left(\frac{n+2}{2}\right)^{\alpha}C_{s,2} \tag{1.2.16}$$

$$P_s < \frac{\pi}{4(n+1)}C_{s,2} \tag{1.2.17}$$

where

$$C_{s,1} = (n+1)n[n(n+1-s)]^{n-s}\frac{2^{n+s}}{(n-s)!}, \tag{1.2.18}$$

$$C_{s,2} = (n+1)^2\left(\frac{3}{2}n+1-s\right)^n\frac{2^n\beta^s}{[(n-s)!(n+1-s)^s]}, \tag{1.2.19}$$

where β is the mesh ratio (see (1.2.11)), α is the exponent in the Lipschitz condition in Theorem 1.2.3.

Th.1.2.3, Th 1.2.4 and Th. 1.2.5 can be found in [C1].

Corollary   If n is fixed, then $\mathcal{L}(f)$ converges uniformly to f on Γ as $|\triangle| \to 0$.

We now present another quasi-interpolant. Since in (1.2.1) and (1.2.2), the coefficients $\{L_j g\}_1^K$ depend on the r-th $(0 \le r \le n)$ derivatives of g at point $t_j \in I_{j,j+n+1}$. We now restrict to the case when the linear functionals in (1.2.1) depend only on the value of g at some point σ_j.

We denote the approximation operator by $S_n^{\alpha,\beta}$

$$(S_n^{\alpha,\beta}g)(z) := \sum_{j=1}^{K} A_j g(\sigma_j)N_{j,n}(z), \quad z \in \Gamma \tag{1.2.20}$$

where

$$\sigma_j = \left(\frac{z_j^{(\beta)}}{z_j^{(\alpha)}}\right)^{\frac{1}{(\beta-\alpha)}}, \quad A_j = (z_j^{(\alpha)})^{\beta/(\beta-\alpha)}(z_j^{(\beta)})^{\alpha/(\alpha-\beta)}$$

, the definition of $z_j^{(\alpha)}, z_j^{(\beta)}$ are given in §1.1, [P4] and α, β are elements in $\{0, \cdots, n\}$.

When $\alpha = 0, \beta = 1$, $S_n^{0,1}$ reproduces linear functions and in this case

$$\sigma_j = \frac{1}{n}(z_{j+1} + \cdots + z_{j+n}), \quad A_j = 1 \quad (j = 1, \cdots, n).$$

For general $\alpha, \beta, 0 \le \alpha, \beta \le n$, we have

$$z^\alpha = \sum_{j=1}^{K} A_j \sigma_j^\alpha N_j^n(z), \tag{1.2.21}$$

$$z^\beta = \sum_{j=1}^{K} A_j \sigma_j^\beta N_j^n(z). \tag{1.2.22}$$

As for the approximation to the function g, we have

Theorem 1.2.6 Let $\rho_n(\underset{\sim}{t}) = \max\{|t_{j+n+1} - t_j| : j = 1, \cdots, K\}$, where $\underset{\sim}{t} = \{t_j\}$ is defined as in (1.1.18). If g is continuous on some neighborhood of Γ, then for fixed α, β as $n\rho_n(\underset{\sim}{t}) \to 0$

$$|(S_n^{\alpha,\beta}g)(z) - g(z)| \le Cn\omega(z^{-\alpha}g(z); \rho_n(\underset{\sim}{t}))$$

where C is a constant and ω denotes the modulus of continuity.

For the approximation to the derivative $g^{(\nu)}(z)$ we have the following theorem.

Theorem 1.2.7 Fix non-negative integers α, β and ν. Suppose g is defined on some neighborhood of Γ and for all η sufficiently close to 1, the function $g(\eta z)$ $(z \in \Gamma)$ lies in $C^\nu(\Gamma)$. Moreover let $h_\nu(\eta z) := \dfrac{d^\nu g(\eta z)}{dz^\nu}$ be a continuous function for z in Γ and η close to 1. Then as $kn^{-2} \to \infty, n \ge \max(\alpha, \beta, \nu + 1)$, there is a constant C such that for $z \in \Gamma$.

$$|(S_n^{\alpha,\beta}g)^{(\nu)}(z) - g^{(\nu)}(z)| \le Cn\left\{\frac{n}{K}\|g^{(\nu)}\| + \omega\left(h_\nu; \frac{n}{K}\right)\right\}$$

where

$$\|g^{(\nu)}\| = \max\{|g^{(\nu)}(z)| : z \in \Gamma\},$$

and the knots $z_1, \cdots, z_k$ are uniformly spaced on Γ.
(cf. [GL1]).

§1.3 Complex Harmonic Splines and Their Function Theoretical Properties. (see [C4] and p.52 Note 4).

Let D be a simply connected domain with boundary γ. A real function $u \in C^{(2)}(D)$ is said to be harmonic in D if it satisfies Laplace's equation $\triangle u = 0$ there. We call a function to be a complex harmonic function on D if it is a finite complex linear combination of harmonic functions in D, and denote the set of all such functions by $H(D)$. These complex harmonic functions share with the (real) harmonic functions many properties, such as the mean value theorem, the principle of maximum modulus, Poisson formula, the Schwartz's theorem (cf. [T]), etc..

Let $\zeta = e^{i\theta}, z \in \overline{U}$. The complex harmonic function

$$P(z) := \frac{1}{2\pi} \int_0^{2\pi} p(\zeta) Re\left(\frac{\zeta + z}{\zeta - z}\right) d\theta \qquad (1.3.1)$$

is said to be a CHSF if $p(\zeta)$ belongs to $\Phi_n(\triangle)$ (refer to §1.1). The family of such functions is denoted by $HS(U)$.

In general, it is difficult to calculate the right side of (1.3.1), but it will be easier if it can be represented by elementary functions. Since the set of boundary values of a CHSF is a complex spline function, therefore, we can decompose any CHSF into a combination of elementary functions. This is one of the strong points of CHSF, we now introduce it in the following.

Let $z = x + iy$, $z_j = x_j + iy_j$. Take

$$f_1(z,j) = [(x_j-x)(x_{j+1}-x)+(y_j-y)(y_{j+1}-y)]|z-z_j|^{-1}|z-z_{j+1}|^{-1} \quad (1.3.2)$$

$$f_2(z,j) = [(|z|^2 x_j - x)(|z|^2 x_{j+1} - x) + (|z|^2 y_j - y)(|z|^2 y_{j+1} - y)]$$
$$\cdot ||z|^2 z_j - z|^{-1} ||z|^2 z_{j+1} - z|^{-1}$$

$$(1.3.3)$$

Define

$$\varphi_{j,j+1}(z) = \cos^{-1} f_1(z,j), \quad \psi_{j,j+1}(\hat{z}) = \cos^{-1} f_2(z,j),$$

where $\hat{z} = (\bar{z})^{-1}, \bar{z}$ is the complex conjugate of $z, \varphi_{j,j+1}(z)$ is the measure of angle from vector $\vec{zz_j}$ to $\vec{zz_{j+1}}$ and $\psi_{j,j+1}(\hat{z})$ is the measure of angle from vector $\vec{\hat{z}z_j}$ to $\vec{\hat{z}z_{j+1}}$.

Theorem 1.3.1 Let $P(z)$ be a CHSF with boundary function $p(\zeta)$ (see (1.3.1)), $p(\zeta) = p_j(\zeta)$ if $\zeta \in \gamma_j$. Then $P(z)$ can be decomposed into

$$P(z) = P_1(z) - P_2(z) \tag{1.3.4}$$

where

$$P_1(z) = -\frac{1}{2\pi i}\sum_{j=1}^{K}(p_j(z) - p_{j-1}(z))\ln|z_j - z| + \frac{1}{2\pi}\sum_{j=1}^{K}p_j(z)\varphi_{j,j+1}(z)$$

$$P_2(z) = -\frac{1}{2\pi i}\sum_{j=1}^{K}(p_j(\hat{z}) - p_{j-1}(\hat{z}))\ln|z_j - \hat{z}| + \frac{1}{2\pi}\sum_{j=1}^{K}p_j(\hat{z})\psi_{j,j+1}(\hat{z}). \tag{1.3.5}$$

Proof.

$$P(z) = \frac{1}{2\pi}\int_0^{2\pi} p(\zeta)\text{Re}\left(\frac{\zeta+z}{\zeta-z}\right)d\theta$$

$$= \frac{1}{4\pi}\int_0^{2\pi} p(\zeta)\frac{\zeta+z}{\zeta-z}d\theta + \frac{1}{4\pi}\int_0^{2\pi}p(\zeta)\frac{\bar{\zeta}+\bar{z}}{\bar{\zeta}-\bar{z}}d\theta$$

$$= \frac{1}{4\pi i}\int_\Gamma\frac{p(\zeta)d\zeta}{\zeta-z} + \frac{z}{4\pi i}\int_\Gamma\frac{p(\zeta)d\zeta}{\zeta(\zeta-z)} + \frac{1}{4\pi i}\int_\Gamma\frac{p(\zeta)}{\zeta}\frac{\bar{\zeta}+\bar{z}}{\bar{\zeta}-\bar{z}}d\zeta$$

$$= \frac{1}{2\pi i}\int_\Gamma\frac{p(\zeta)}{\zeta-z}d\zeta - \frac{1}{4\pi i}\int_\Gamma\frac{p(\zeta)}{\zeta}d\zeta + \frac{1}{4\pi i}\int_\Gamma p(\zeta)\left(\frac{1}{\zeta} - \frac{2}{\zeta-\hat{z}}\right)d\zeta$$

$$= \frac{1}{2\pi i}\int_\Gamma\frac{p(\zeta)}{\zeta-z}d\zeta - \frac{1}{2\pi i}\int_\Gamma\frac{p(\zeta)}{\zeta-\hat{z}}d\zeta$$

$$= P_1(z) - P_2(z). \tag{1.3.6}$$

Since

$$p(\zeta) = p_j(\zeta) = \sum_{k=0}^{n}a_k^{(j)}(\zeta-z)^k \in \pi_n$$

when $\zeta \in \gamma_j$, set $\sigma_k^{(j-1)} = a_k^{(j)} - a_k^{(j-1)}$; and $\sigma_k^{(0)} = \sigma_k^{(K)} = a_k^{(1)} - a_k^{(K)}$, for $k = 0, \cdots, n$. By induction, we have

$$\sigma_{n-k}^{(j-1)} = (-1)^k\binom{n}{k}\sigma_n^{(j-1)}(z_j - z)^k, \quad k = 0, \cdots, n \tag{1.3.7}$$

$$\sum_{j=1}^{K} \sigma_n^{(j-1)}(z_j - z)^k = 0, \quad k = 0, \cdots, n. \tag{1.3.8}$$

By using (1.3.7) and (1.3.8), we have

$$P_1(z) = \frac{1}{2\pi i} \int_\Gamma \frac{p(\zeta)d\zeta}{\zeta - z} = \frac{1}{2\pi i} \sum_{j=1}^{K} p_j(z) \int_{z_j}^{z_{j+1}} \frac{d\zeta}{\zeta - z}, \tag{1.3.9}$$

$$P_2(z) = -\frac{1}{2\pi i} \int_\Gamma \frac{p(\zeta)d\zeta}{\zeta - \hat{z}} = -\frac{1}{2\pi i} \sum_{j=1}^{K} p_j(\hat{z}) \int_{z_j}^{z_{j+1}} \frac{d\zeta}{\zeta - \hat{z}}. \tag{1.3.10}$$

Thus (1.3.5) follows from (1.3.9) and (1.3.10).   □.

CHSF has a very nice approximation property (see also p.55 Note 4):

Suppose that a complex harmonic function G, of which we only know its values at some points on Γ, is to be recovered. Then we can construct interpolant or quasi-interpolant $p(\zeta)$ which fits G at these points. Taking $P(z)$ with boundary function $p(\zeta)$, then $P(z)$ approximates $G(z)$ very well on $\overline{U}$ if the mesh points of $p(\zeta)$ are close to each other. It is surprising that not only the functional properties but also the geometric shape of both $P(z)$ and $G(z)$ are very close.

Now we introduce some spaces of functions.

$C^{(n)}(\overline{D}) = \{f |$ for any $s, 0 \le s \le n$, the s-th derivative of f

with respect to variable z is continuous on $\overline{D}\}$,

$C(\overline{D}) \quad = C^{(0)}(\overline{D}).$

$H^{(n)}(\overline{D}) = C^{(n)}(\overline{D}) \cap H(D),$

$AC^{(n)}(\Gamma) = \{f |$ the n-th derivative of f is absolutely

continuous on $\Gamma\}$,

$AH^{(n)}(\overline{D}) = AC^{(n)}(\Gamma) \cap H^{(n)}(\overline{D}), \quad \Gamma = \partial U$ is the unit circle.

Theorem 1.3.2 Let F be analytic in U, $F \in AH^{(n)}(\overline{U}), n \ge 2$, and let $p = \mathcal{L}(F)$ be the quasi-interpolant (see (1.2.1)) complex spline function.

Then the CHSF

$$P(z) = \frac{1}{2\pi} \int_0^{2\pi} p(\zeta) \text{Re} \left(\frac{\zeta+z}{\zeta-z} \right) d\theta, \quad \zeta = e^{i\theta}, z \in U \qquad (1.3.11)$$

approximates F in U as follows:

$$|F(z) - P(z)| \le \frac{K_0}{2\pi} \omega(F^{(n)}; |\triangle|)|\triangle|^n, \quad z \in \overline{U}, \qquad (1.3.12)$$

$$|F'(z) - P_z(z)| \le \left(\frac{\pi}{2} K_2 + K_1 |\triangle| \right) \omega(F^{(n)}; |\triangle|)|\triangle|^{n-2}, \quad z \in \overline{U}, \ (1.3.13)$$

$$|P_{\overline{z}}(z)| = |F_{\overline{z}}(z) - P_{\overline{z}}(z)|$$

$$\le \frac{\pi}{2}(K_2 + K_1 |\triangle|)\omega(F^{(n)}; |\triangle|)|\triangle|^{n-2}, \quad z \in \overline{U}, \ (1.3.14)$$

where $\omega(f : |\triangle|)$ is the modulus of continuity of f on Γ, and K_0, K_1 and K_3 are constants given in (1.2.8).

This theorem tells us that if $|\triangle|$ tends to zero, the function P and its derivatives converge uniformly to F and its derivatives respectively on the closed disc $|z| \le 1$. This is another advantage of using CHSF, which will become more clear later.

The proof of Theorem 1.3.2 goes as follows.

Using the maximum modulus principle and (1.2.8), we have (1.3.12). Now

$$|P_z(z) - F'(z)| \le \sup_{z_0 \in \Gamma} \left| \frac{1}{2\pi} \int_\Gamma \frac{\varphi(\zeta, z_0)}{\zeta - z} d\zeta + p'(z_0) - F'(z_0) \right|, \quad z \in \overline{U}.$$

where

$$\varphi(\zeta, z_0) = (p'(\zeta) - F'(\zeta)) - (p'(z_0) - F'(z_0))$$

$$|\varphi(\zeta, z_0)| = \left| \int_{\gamma(z_0, \zeta)} (p^{(2)}(t) - F^{(2)}(t)) dt \right|$$

$$\le \max_{t \in \Gamma} |p^{(2)}(t) - F^{(2)}(t)||\gamma(z_0, \zeta)|$$

where $|\gamma(z_0, \zeta)|$ is the arc length of $\gamma(z_0, \tau)$. Hence we have

$$2 \int_{\tilde{\Gamma}_1} \left| \frac{\varphi(\zeta, z_0)}{\zeta - z_0} \right| |d\zeta| \ge \left| \int_\Gamma \frac{\varphi(\zeta, z_0)}{\zeta - z_0} d\zeta \right|$$

where $\tilde{\Gamma}_1$ is one of the half circles $\tilde{\gamma}_1$ and $\tilde{\gamma}_2$ such that

$$\left| \int_{\tilde{\Gamma}_1} \frac{\varphi(\zeta, z_0)}{\zeta - z_0} d\zeta \right| = \max \left\{ \left| \int_{\tilde{\gamma}_1} \frac{\varphi(\zeta, z_0)}{\zeta - z_0} d\zeta \right|, \left| \int_{\tilde{\gamma}_2} \frac{\varphi(\zeta, z_0)}{\zeta - z_0} d\zeta \right| \right\}$$

$$\tilde{\gamma}_1 = \{z \mid z = e^{i\theta}, \quad 0 \leq \theta \leq \pi\},$$

$$\tilde{\gamma}_2 = \{z \mid z = e^{i\theta}, \quad \pi \leq \theta \leq 2\pi\},$$

where we stipulate that $|\gamma(z_0, \zeta)| \leq \pi$. Since $1 \leq |\gamma(z_0, \zeta)|/|z_0 - \zeta| \leq \pi/2$,

$$2 \int_{\tilde{\Gamma}_1} \frac{|\varphi(\zeta, z_0)|}{|\zeta - z_0|} |d\zeta| \;\leq\; \pi^2 \max_{t \in \Gamma} |p^{(2)}(t) - F^{(2)}(t)|$$

$$|F'(z) - P_z(z)| \;\leq\; \frac{\pi}{2} \max_{t \in \Gamma} \left| F^{(2)}(t) - p^{(2)}(t) \right| + \max_{t \in \Gamma} \left| F'(t) - p'(t) \right|,$$

then from (1.2.8), we have (1.3.13).

Since

$$|F_{\bar{z}}(z) - P_{\bar{z}}(z)| \leq \frac{1}{2\pi} \sup_{z_0 \in \Gamma} \left| \int_\Gamma \frac{\psi(\zeta, z_0)}{\zeta - z_0} d\zeta \right|,$$

where

$$\psi(\zeta, z_0) = (\zeta p'(\zeta) - F'(\zeta)) - (z_0 p'(z_0) - z_0 F'(z_0)),$$

$$|\psi(\zeta, z_0)| \leq \max_{\zeta \in \Gamma} \{ |p'(\zeta) - F'(\zeta)| + |p^{(2)}(\zeta) - F^{(2)}(\zeta)| \} |\gamma(z_0, \zeta)|,$$

we have

$$|P_{\bar{z}}(z)| \leq \frac{\pi}{2} \max_{\zeta \in \Gamma} \{ |p'(\zeta) - F'(\zeta)| + |p^{(2)}(\zeta) - F^{(2)}(\zeta)| \}.$$

from (1.2.8), we obtain (1.3.14). $\square$.

Comment   Results similar to those in Theorem 1.3.2, Theorem 1.3.3 and Theorem 1.3.4 can be obtained when $\mathcal{L}(F)$ is replaced by $I(F)$–the interpolating complex spline as introduced in §1.1.

Corollary 1.3.1   Let $F(z)$ and $P(z)$ be defined as in Theorem 1.3.2. Then

$$\left| \frac{\partial^j P(z)}{\partial z^j} - F^{(j)}(z) \right| \leq \left[\frac{\pi}{2} K_{j+1} + K_j |\Delta| \right] \omega(F^{(n)}; |\Delta|) |\Delta|^{n-j-1},$$

where $j = 1, \cdots, n-1, z \in \overline{U}$. K_j, K_{j+1} are constants given in (1.2.8).

The following theorem gives conditions under which the complex harmonic spline is an open mapping.

Theorem 1.3.3　Let F be analytic in U, $F \in AH^{(n)}(\overline{U})$, $n \geq 2$. Assume that $F'(z) \neq 0$, $z \in \overline{U}$. Choose $|\triangle|$ sufficiently small such that

$$
\begin{aligned}
\omega \;&<\; \min_{z \in \Gamma} |F'(z)|/\pi(K_2 + 2K_1), \quad n = 2 \\[2mm]
\omega^{1/(n-1)} |\triangle| \;&<\; \left[\min_{z \in \Gamma} |F'(z)|/\pi(K_2 + 2K_1) \right]^{1/(n-2)}, \quad n > 2
\end{aligned}
\tag{1.3.15}
$$

where $\omega = \omega(F^{(n)}; |\triangle|)$ is the modulus of continuity of $F^{(n)}$ on Γ. Then the Jacobian J of CHSF P with boundary $p = \mathcal{L}(F)$ is positive on $\overline{U}$.

Proof.　Let

$$
\eta = \left(\frac{\pi}{2} K_2 + K_1 |\triangle| \right) \omega |\triangle|^{n-2}, \quad \xi = \frac{\pi}{2}(K_2 + K_1 |\triangle|)\, \omega |\triangle|^{n-2}. \tag{1.3.16}
$$

Since $|\triangle| < 2$, from (1.3.13) and (1.3.14), we have

$$
J(z) = |P_z|^2 - |P_{\bar{z}}|^2 \geq (|F'(z)| - \eta)^2 - \xi^2.
$$

From (1.3.15), we obtain $J(z) > 0$.　　　　　　　　　　　　　　　　□.

It is easy to prove the following:

Lemma 1.3.1　Let γ be a closed Jordan curve. γ is a homomorphic image of Γ, $\gamma = f(\Gamma)$, $f \in C^{(1)}(\Gamma)$. If $f'(z) \neq 0$ for $z \in \Gamma$, then

$$
\begin{aligned}
m_f &:= \inf_{z_1, z_2 \in \Gamma} \left| \frac{f(z_1) - f(z_2)}{z_1 - z_2} \right| > 0, \\[2mm]
M_f &:= \sup_{z_1, z_2 \in \Gamma} \left| \frac{f(z_1) - f(z_2)}{z_1 - z_2} \right| < \infty.
\end{aligned}
\tag{1.3.17}
$$

Theorem 1.3.4　Let D be a simply connected domain, $\partial D = \gamma$ a closed Jordan curve with bounded curvature, $W = F(z)$ a conformal mapping of U onto D, and $F \in AH^{(n)}(\overline{U})$, $n \geq 2$. If $|\triangle|$ is so small that

$$
\begin{aligned}
\omega \;&<\; m_f/\pi(K_2 + 2K_1), \quad n = 2, \\[2mm]
\omega^{1/(n-2)} |\triangle| \;&<\; (m_F/\pi(K_2 + 2K_1))^{1/(n-2)}, \quad n > 2
\end{aligned}
\tag{1.3.18}
$$

where $\omega = \omega(F^{(n)}; |\Delta|), m_F$ defined as in (1.3.17), then the function

$$P(z) = \frac{1}{2\pi} \int_0^{2\pi} \mathcal{L}(F) \text{Re}\left(\frac{\zeta+z}{\zeta-z}\right) d\theta$$

maps U onto a simply connected domain D_p. This mapping is $1-1$ and sensepreserving; moreover,

$$\lim_{|\Delta|\to 0} D_p = D \tag{1.3.19}$$

in the sense of Caratheodory.

Proof. Since $F \in C^{(n)}(\overline{U}), \gamma = F(\Gamma)$ is smooth, $\arg F'(\zeta)$ is continuous in $\overline{U}$ and $\arg F'(\zeta) = \varphi(\zeta) - \arg\zeta - \pi/2$, where $\varphi(\zeta) = \arg(dF(e^{i\theta})/d\theta)$. Let S denote the arc length of γ. Then

$$|S_2 - S_1| = \int_{\theta_1}^{\theta_2} |F'(e^{i\theta})| d\theta$$

$$\leq \left(\int_{\theta_1}^{\theta_2} |F'(e^{i\theta})|^2 d\theta\right)^{1/2} \left(\int_{\theta_1}^{\theta_2} d\theta\right)^{1/2}$$

$$\leq K_1(\theta_2 - \theta_1)^{1/2}.$$

Let $K(\zeta)$ be the curvature of γ at point ζ. From the hypothesis, $K(\zeta) \leq K_0 < \infty$, where K_0 is a constant, hence

$$|\arg F'(\zeta_2) - \arg F'(\zeta_1)| = \left|\int_{S_1}^{S_2} \frac{d\varphi}{ds} ds + (\theta_1 - \theta_2)\right|$$

$$\leq K_0|s_2 - s_1| + |\theta_2 - \theta_1|$$

$$\leq K_0 K_1|\theta_2 - \theta_1|^{1/2} + |\theta_2 - \theta_1|$$

$$\leq K|\theta_2 - \theta_1|^{1/2}.$$

Since $F'(z) \neq 0$ for $z \in U, \ln F'$ is analytic in U and there exists a constant K' such that

$$|\ln F'(z_2) - \ln F'(z_1)| \leq K'|z_1 - z_2|^{1/2}, \quad z_1, z_2 \in \overline{U}$$

(cf. [Go, Chap 9, Set 5, Th.4,5]). Therefore $\ln F'$ is continuous and hence bounded in $\overline{U}$. We then conclude that $F'(z) \neq 0, z \in \Gamma$; hence from Lemma 1.3.1, we have $m_F > 0$.

The directional derivative of a complex harmonic function G can be written as $\partial G/\partial l_\theta = G_z e^{i\theta} + G_{\bar{z}} e^{-i\theta}$, and this is a complex harmonic function.

From Theorem 1.3.2, we obtain

$$\sup_{z \in \overline{U}} \left| \frac{\partial P}{\partial \theta} - \frac{\partial F}{\partial \theta} \right| \leq \sup_{z \in \Gamma} |P_z - F'| + \sup_{z \in \Gamma} |P_{\bar{z}}|$$
$$\leq \eta + \xi$$

where η and ξ are defined by (1.3.16).

From (1.3.13), (1.3.14) we obtain

$$\left| \frac{P(z_2) - P(z_1)}{z_2 - z_1} - \frac{F(z_2) - F(z_1)}{z_2 - z_1} \right| \leq \eta + \xi;$$

therefore

$$|z_2 - z_1|(m_F - \eta - \xi) \leq |P(z_2) - P(z_1)|$$
$$\leq (M_F + \eta + \xi)|z_2 - z_1|, \tag{1.3.20}$$

from (1.3.18), $\eta + \xi < m_F$. We conclude that

$$P(z_2) = P(z_1) \quad \text{if and only if} \quad z_2 = z_1,$$

i.e., $w = P(z)$ is a homeomorphism. Note that

$$m_F \leq \min_{z \in \Gamma} |F'(z)|. \tag{1.3.21}$$

From (1.3.21) and (1.3.18) we obtain inequalities (1.3.15). Theorem 1.3.3 tells us

$$J(z) > 0 \quad \text{for} \quad z \in \overline{U}.$$

The mapping $w = P(z)$ from U onto D_p is sensepreserving (cf. [Go]). Since the Jacobian is positive for $z \in \overline{U}$, the mapping $w = P(z)$ is open; thus $D_p = P(U)$ is a domain, and no interior point of U can be mapped onto the boundary of D_p. Hence the boundary of D_p must be the image of Γ. Since (1.3.20) is valid on $\overline{U}, P$ is homeomorphic from Γ to $\gamma = \partial D_p$. We now conclude that γ is a closed Jordan curve and D_p is a simply connected domain.

If $z \in \Gamma$, from Theorem 1.3.2 we have

$$|P(z) - F(z)| < \frac{K_0}{2\pi}\omega(F^{(n)}; |\triangle|)| \triangle |^n;$$

therefore (1.3.19) is proved. $\square$.

Since $P_z, P_{\overline{z}}$ are continuous in $\overline{U}$, denote the complex dilatation by $\chi(z) := P_{\overline{z}}(z)/P_z(z), z \in \overline{U}$, and

$$D(z) = \frac{|P_z(z)| + |P_{\overline{z}}(z)|}{|P_z(z)| - |P_{\overline{z}}(z)|}.$$

We can prove the following

Theorem 1.3.5   Let F be a conformal mapping of U onto $D, F \in AH^{(n)}(\overline{U})$. Let P be a CHSF as defined in (1.3.11). Choose $|\triangle|$ so small that

$$\lambda - \pi(K_2 + 2K_1)\omega(F^{(n)}; |\triangle|)| \triangle |^{n-2}$$
$$< m_F, \quad n \geq 2. \tag{1.3.22}$$

Let ε be any positive number satisfying the relation

$$\lambda < \varepsilon \leq m_F. \tag{1.3.23}$$

Then P satisfies the Beltrami differential equation

$$W_{\overline{z}} = \chi W_z \tag{1.3.24}$$

where χ is continuous in $\overline{U}$. Further, P is a K-quasiconformal mapping with dilatation

$$D(z) < K = \frac{M + \varepsilon}{M - \varepsilon}, \quad |\chi(z)| < \frac{\varepsilon}{M} \tag{1.3.25}$$

where

$$M := \sup_{z \in \Gamma} |F'(z)|.$$

Proof. From (1.3.12), (1.3.13)

$$|D(z)| \leq \frac{|F'(z)| + \eta + \xi}{|F'(z)| - \eta - \xi},$$

with η, ξ defined by (1.3.16). From (1.3.22), (1.3.23)

$$|D(z)| < \frac{M+\lambda}{M-\lambda}$$

$$< \frac{M+\varepsilon}{M-\varepsilon} = K$$

and

$$|\chi(z)| < \frac{K-1}{K+1}$$

$$= \frac{\varepsilon}{M}.$$

Then (1.3.25) is proved.

Now $\lambda < m_F$, (1.3.18) is valid, from Theorem 1.3.4, $w = P(z)$ is a homeomorphism and the function P satisfies (1.3.24); obviously, $\chi(z)$ is continuous in $\overline{U}$.

$\square$.

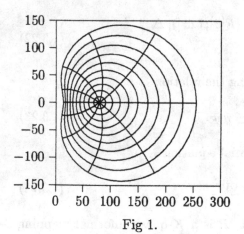

Fig 1.

Figure 1 is a set of curves which are the image of radii and concentric circles in the disc $\overline{U}$ under the mapping

$$w = P(z) = \frac{1}{2\pi} \int_0^{2\pi} p(\zeta) \text{Re} \left(\frac{\zeta+z}{\zeta-z} \right) d\theta, \quad \zeta = e^{i\theta}, \quad |z| \leq 1,$$

where p is the complex cubic spline which interpolates $F(z)$ at points: $z_j = e^{i(2\pi/20)}$, $j = 1, \cdots, 20$, where $F(z) = (z-3)^4 + 8/(z-5) + 12/(z-5)^2$, i.e., $p(z_j) = F(z_j)$, $j = 1, \cdots, 20$.

We see that the two families of curves are almost perpendicular. The reason is that since $w = F(z)$ is a conformal mapping on $\overline{U}$, then $w = P(z)$ is almost a conformal mapping on $\overline{U}$. The rigorous proof is contained in Theorems 1.3.2–1.3.4.

§1.4 Geometric Property of CHSF

Since the conformal mapping is 1–1 and it maps an open set onto an open set. We would ask: does the CHSF also keep these properties if the mesh size being sufficiently small?

We will answer these questions in the following theorems (refer to [C11]).

Theorem 1.4.1   Let P be a CHSF,

$$P(z) = \frac{1}{2\pi} \int_0^{2\pi} p(\zeta) \mathrm{Re}\left(\frac{\zeta + z}{\zeta - z}\right) d\theta$$

, $\zeta = e^{i\theta}, z \in U; P(U)$ is the image of unit disc U under the mapping $P : z \to w$, then

$$P = f(\phi) \quad (= f \circ \phi) \tag{1.4.1}$$

where $\phi : z \to \phi(z)$ is a homeomorphism from U onto a domain Ω; f is analytic on Ω. $\phi(t)$ is in $C^\infty(U)$ possibly except on a discrete set. And $\phi(z)$ is harmonic iff one of the following conditions is satisfied:

(i) $f(\phi) = A\phi + B$, A, B are constants,
(ii) $P(z)$ is a polynomial of degree n.

Now if we add some conditions on $P(z)$ along Γ, then the conclusion will be strengthened.

Theorem 1.4.2   Let P be a CHSF, which is non-constant on any subarc of Γ. Then P has the form as (1.4.1), where ϕ is a homeomorphism and maps the closed unit disc $\overline{U}$ onto a closed unit disc $\overline{\Omega}$, f is analytic in Ω and continuous on $\overline{\Omega}$.

A CHSF is a harmonic function, which in general cannot be a conformal mapping. However, under some conditions it could be a composition of a homeomorphism and a conformal mapping. In fact, we have:

Theorem 1.4.3   Let $P(z)$ be a function satisfying the condition in Theorem 1.4.2. If $\gamma = P(\Gamma)$ is a Jordan curve, then $P : U \mapsto P(U)$ is a $1 - 1$ mapping, the function f in (1.4.1) is a conformal mapping, and the two closed domains $\overline{\Omega}$ and $\overline{f}(\Omega)$ are homeomorphic.

If a point Q is covered by the image $P(U)$ we want to know the covering number, is it finite or infinite? The following theorem and the corollary give an upper bound on the covering number.

Theorem 1.4.4   Suppose $P(z)$ satisfies the conditions in Theorem 1.4.2, and $P(\zeta) \neq 0$ on Γ. Let N denote the number of zeros of $P(z)$. n and K are the degree and number of knots of the spline function $p(\zeta)$ respectively. Then

$$N < nK. \tag{1.4.2}$$

Corollary 1.4.1   Suppose $P(z)$ satisfies the condition in Theorem 1.4.2, Q is a point in the image $P(U)$, and N denotes the covering number of Q. Then

$$N < nK \tag{1.4.3}$$

where n and K are the same as in Theorem 1.4.4.

Corollary 1.4.2   Let $F(z)$ and $\Phi(z)$ be two CHSF satisfying the conditions in Theorem 1.4.2, and $|F(z)| > |\Phi(z)|$ for $z \in \Gamma$. Then $F(z) + \Phi(z)$ and $F(z)$ have the same number of zeros in U.

Corollary 1.4.3   Let γ be a closed Jordan curve in $\overline{U}$, Ω_γ the domain bounded by γ. $P_k(z)$ and $P(z)$ are CHSF satisfying the conditions in Theorem 1.4.2, $P_k(z)$ converge uniformly to $P(z)$ on Ω_γ. Suppose $P(z) \neq 0$

for $z \in \gamma$. Then, there exists an integer ν, if $k > \nu$, $P_k(z)$ and $P(z)$ have the same number of zeros in Ω_γ.

The following corollary provides a method of counting the number of zeros of $P(z)$ in U by comparing the magnitude of the coefficients of $p(\zeta)$.

Corollary 1.4.4   Let $P(z)$ be a CHSF. Its boundary function is $p(\zeta)$ (see (1.3.1)). Assume that

$$p(\zeta) = a_{n,j}\zeta^n + \cdots + a_{1,j}\zeta + a_{0,j}, \quad \zeta \in \gamma(z_j, z_{j+1}),$$

$j = 1, \cdots, K$, $p(\zeta) \in C^{n-1}(\Gamma)$. $p(\zeta)$ is not a constant on each $\gamma_{j,j+1}, j = 1, \cdots, K$. If there exists a_ν which is independent of j, such that

$$|a_{n,j}| + \cdots + |a_{\nu+1,j}| + |a_{\nu-1,j}| + \cdots + |a_{0,j}| < |a_\nu|, \quad j = 1, \cdots, K$$

then $P(z)$ has ν zeros in U.

The proof of the last two corollaries are similar to that for analytic functions (cf. [Mar]).

We now establish some lemmas.

The complex derivatives $\partial = \partial_z, \overline{\partial} = \partial_{\overline{z}}$ are defined as

$$\partial = \frac{1}{2}\left(\frac{\partial}{\partial x} - i\frac{\partial}{\partial y}\right), \quad \overline{\partial} = \frac{1}{2}\left(\frac{\partial}{\partial x} + i\frac{\partial}{\partial y}\right),$$

the Jacobian of $P(z)$ is

$$\begin{aligned} J(z) &= |\partial P(z)|^2 - |\overline{\partial} P(z)|^2 \\ &= |F_1(z)|^2 - |\overline{F}_2(z)|^2; \end{aligned} \tag{1.4.4}$$

where

$$F_1(z) = \frac{1}{2\pi i}\int_\Gamma \frac{p(\zeta)}{(\zeta - z)^2}d\zeta, \quad F_2(z) = \frac{1}{2\pi i}\int_\Gamma \frac{\overline{p}(\zeta)}{(z\zeta - 1)^2}d\zeta \tag{1.4.5}$$

and $\overline{p}(\zeta) = \overline{a}_{n,j}\zeta^n + \cdots + \overline{a}_{0,j}$, $\zeta \in \gamma_j$, $j = 1, \cdots, K$.

We can also represent $F_1, \overline{F}_2$ into Cauchy types of integrals:

$$F_1(z) = \frac{1}{2\pi i}\int_\Gamma \frac{p^{(1)}(\zeta)}{\zeta - z}d\zeta, \quad \overline{F}_2(z) = \frac{\hat{z}^2}{2\pi i}\int_\Gamma \frac{p^{(1)}(\zeta)}{\zeta - \hat{z}}d\zeta \quad (z \neq 0), \tag{1.4.6}$$

where $\hat{z} = \frac{1}{\bar{z}}$.

Set

$$\omega = \{z|J(z) = 0, z \in U\}, \quad \omega_1 = \{z|z \in \omega, F_1(z) = F_2(z) = 0\},$$

$$\omega_0 = \omega\backslash\omega_1.$$

(1.4.7)

Lemma 1.4.1   Let $P(z)$ be a CHSF satisfying the conditions in Theorem 1.4.2. Then ω does not include any open set, moreover, ω_1 has no accumulation points in U. If ω_0 is not empty, then ω_0 includes some isolated points, by adding some analytic curves.

Proof.   Suppose $\xi_j \in \omega_1$ and $\lim\limits_{j\to\infty} \xi_j = \xi \in U$. Since $F_1(\xi_j) = F_2(\xi_j) = 0$, letting j tend to infinity, we have $F_i(\xi) = 0$ $(i = 1, 2)$. Therefore, $F_1(z) = F_2(z) = 0$ for all $z \in U$ since $F_i(z)$ is analytic in D. Thus

$$\hat{z}^2 \frac{1}{2\pi} \int_0^{2\pi} p^{(1)}(\zeta)\text{Re}\left(\frac{\zeta + z}{\zeta - z}\right) d\theta = \hat{z}^2 F_1(z) - \overline{F}_2(z) = 0, \qquad (1.4.8)$$

let $z \to \xi_0 \in \Gamma$. Hence $p^{(1)}(\xi_0) = 0$ for all $\xi_0 \in \Gamma$, $p(\zeta)$ is a constant, and hence $P(z) \equiv \text{const.}$, this is a contradiction to the hypothesis. We conclude that ω_1 has no accumulation points in U.

Suppose on the contrary that ω_0 contains an open set $\tilde{\omega}_0$. Let z be a point in $\omega_0 = \omega\backslash\omega_1$. From (1.4.4), we have $|F_1(z)| + |F_2(z)| \neq 0$ and $|F_1(z)| = |F_2(z)| \neq 0$ for $z \in \tilde{\omega}_0$. Set $G(z) = F_1(z)/F_2(z)$. Then $\text{Re}(\ln G(z)) = 0$, $\arg G(z) = \text{const.}$ for $z \in \omega_0$, and $G(z) = e^{ic}$ for $z \in \tilde{\omega}_0$. Therefore,

$$F_1(z) = e^{ic}F_2(z), \quad \text{for} \quad z \in U \qquad (1.4.9)$$

From (1.4.6),

$$F_1(z) = \sum_{j=0}^{\infty}(j+1)a_j z^j, \quad a_j = \frac{1}{2\pi i}\int_\Gamma p(\zeta)\zeta^{-j+2}d\zeta$$

$$F_2(z) = \sum_{j=0}^{\infty}(j+1)b_j z^j, \quad b_j = \frac{1}{2\pi i}\int_\Gamma \overline{p}(\zeta)\zeta^j d\zeta$$

(1.4.10)

because $p(\zeta) \in C^{n-1}(\Gamma), n \geq 1$, and is piecewise analytic function on Γ. Therefore, the Fourier series of $p(\zeta)$ converges uniformly and absolutely in $[0, 2\pi]$ to $p(\zeta)$.

The Fourier expansion of p is

$$p(e^{i\theta}) = \sum_{m \in Z} C_m e^{im\theta}, \quad C_m = \frac{1}{2\pi} \int_0^{2\pi} p(e^{i\theta}) e^{-im\theta} d\theta,$$

it is easy to prove that

$$\bar{p}(e^{i\theta}) = \sum_{m \in Z} \overline{C}_m e^{im\theta}.$$

From (1.4.9) and (1.4.10), we have $a_j = e^{iC} b_j, j = 0, 1, \cdots$. Therefore,

$$e^{-i/2C} C_{j+1} = e^{i/2C} \overline{C}_{-j-1},$$

thus

$$\begin{aligned} p(e^{i\theta}) &= C_0 + \sum_{j=1}^{\infty} (C_j e^{ij\theta} + e^{iC} C_{-j} e^{-ij\theta}) \\ &= C_0 + e^{iC/2} h(\theta), \end{aligned}$$

where $h(\theta)$ $(0 \le \theta \le 2\pi)$ is a real-valued function. This is impossible, since $p(\zeta)$ is a polynomial of degree n $(n \ge 1)$ on $\gamma_j (j = 1, \cdots, K)$. We thus conclude that the set ω_0 does not contain any open set.

Let z be in ω_0. Since $|F_1(z)| = |F_2(z)| \ne 0$, we define a function $G(z) = F_1(z)/F_2(z)$ as before. Let γ_α be a component of ω_0 and a continuum containing more than one point, i.e., it is a nonempty, bounded, connected closed set, where ω_0 is already defined by (1.4.7). $G(z)$ is analytic on γ_α and the function $G^{(1)}(z)$ could have a finite number of zeros, denoted by $\{A_j\}_1^m$, on γ_α. Thus $G^{(1)}(z_0) \ne 0$ if $z_0 \in \gamma_\alpha \backslash (\bigcup_1^m A_j)$. There is a sufficiently small neighborhood $O(z_0)$ of point z_0. The function $W = G(z)$ maps $O(z_0)$ onto $O(w_0) = G(O(z_0))$, the neighborhood of point w_0. Since $|G(z)| = 1$ for $z \in \gamma_\alpha$, the image of the set $\gamma_{z_0} := O(z_0) \cap \{\gamma_\alpha \backslash (\bigcup_1^m A_j)\}$ is an arc Γ_{w_0} on the unit circle, which is an analytic arc. Because of the analyticity of the inverse function $G^{-1}(w), w \in \Gamma_{w_0}, \gamma_{z_0}$ is also an analytic arc. We conclude that γ_α consists of analytic arcs. □

Lemma 1.4.2 Let $P(z)$ be a CHSF satisfying the condition in Theorem 1.4.2, and $n \ge 1$. Then, except at most finite points, we have

$$J(z) \ne 0, \quad z \in U. \tag{1.4.11}$$

There are no accumulation points of the zeros of $J(z)$ in U.

Proof.   Suppose that $p(\zeta)$ is a polynomial of degree $n, n \geq 1$. Since the complex harmonic function is determined uniquely by its boundary values, $P(z)$ is also a polynomial of degree n. $J(z) = |P^{(1)}(z)|^2$, and $P^{(1)}(z)$ is a polynomial of degree $n-1$, having at most $n-1$ zeros in U. Consequently, hereafter we consider only the case that $p(\zeta)$ is a spline of degree n, not a polynomial. We conclude that the zeros of the function $F_2(z)$ have no limit points in U. If this were not true, it would be $F_2(z) = 0$ for all $z \in U$. From (1.4.10), $b_j = 0, j = 0, 1, \cdots$. Thus $\int_\Gamma \overline{p}(\zeta)\zeta^j d\zeta = 0$. It is easy to derive $\int_\Gamma p(\zeta)\zeta^j d\zeta = 0$ for $j = 1, 2, \cdots$. By integration by parts, we have

$$\sum_{\nu=1}^K \int_{\gamma_\nu} \zeta^{j+n} p^{(n)}(\zeta)d\zeta = 0, j = 0, 1, \cdots.$$

Set $\sigma_n^{(\nu-1)} = a_{n,\nu} - a_{n,\nu-1}$. Then

$$\sum_{\nu=1}^K \sigma_n^{(\nu-1)} z_\nu^{j+n+1} = 0, \ j = 0, 1, \cdots.$$

Using (1.3.8) we obtain

$$\sum_{\nu=1}^K \sigma_n^{(\nu-1)} z_\nu^j = 0, j = 0, 1, \cdots.$$

Since $z_1, z_2, \cdots, z_K$ are distinct, $\sigma_n^{(\nu-1)} = 0, \nu = 1, \cdots, K$, that is $a_{n,1} = a_{n,2} = \cdots = a_{n,K}(=: a_{n,0})$, but $p(\zeta) \in C^{n-1}(\Gamma)$, we conclude that $p(\zeta)$ is a polynomial, contradicting our assumption. Thus the zeros of $F_2(z)$ have no limit point in U.

Let γ_α be an analytic arc in ω_0. Set $G(z) = F_1(z)/F_2(z)$ for $z \in \gamma_\nu$. We have proved in Lemma 1.4.1, that the function $G^{(1)}(z)$ has at most finite number of zeros on γ_α. Without loss of generality we assume that $G^{(1)}(z) \neq 0$ for $z \in \gamma_\alpha$ and γ_α is a Jordan arc. Since the arc γ_α is analytic let z_0 be a point on γ_α. There is an interval $I_\delta = (t_0 - \delta, t_0 + \delta) \subset [0, 1]$ such that $\gamma_\alpha = \{z | z = z(t), t \in [0, 1]\}$ and the arc $\gamma_\alpha \cap O(z_0)$ can be represented as

$$z = z(t) = \sum_{m=0}^\infty C_m(t - t_0)^m, z'(t) \neq 0$$

for $t \in I_\delta$, where $O(z_0)$ is a neighborhood of z_0. By Abel's theorem the above series also converges in the disc $U_\delta(t_0) = \{t \big| |t - t_0| < \delta\}$, thus $z(t)$ is an analytic function in $U_\delta(t_0)$. Without loss of generality, we may regard the image of $U_\delta(t_0)$ under the mapping $z : t \mapsto z(t)$ is $O(z_0)$. $O(z_0)$ is cut by γ_α into two parts, the upper half disc $U_\delta^+(t_0)$ corresponding to $O^+(z_0)$ and the lower half disc $U_\delta^-(t_0)$ to $O^-(t_0)$. Set $z^* = z(\bar{t})$ and define a function $F(z)$:

$$F(z) = \begin{cases} G(z), & z \in O^+(z_0) \cup \gamma_\alpha, \\ \dfrac{1}{\overline{G(z^*)}}, & z \in O^-(z_0). \end{cases}$$

By the symmetry principle [LS, Ch.2, Sec 3], $F(z)$ is an analytic function on $O(z_0)$. $G(z)$ is analytic on γ_α, and also on $O(z_0)$ for sufficiently small $O(z_0)$. From the theorem of uniqueness of analytic continuation, we conclude that $F(z) = G(z)$ for $z \in O(z_0)$. Thus $G(z) = 1/\overline{G}(z^*)$ for $z \in O(z_0)$. By (1.4.6), a simple calculation leads to $G(z) = G(1/z^*)$ for $z \in O(z_0)$. Since $G^{(1)}(z_0) \neq 0$, the inverse function G^{-1} exists. Therefore, $z = G^{-1}G(1/z^*)$. In particular, letting $z = z_0$, we obtain $z_0 = 1/\bar{z}_0$, thus $|z_0| = 1$. This contradicts the fact that $z_0 \in \gamma_\alpha \subset U$. The above argument shows that ω_0 does not contain any analytic arcs. By Lemma 1.4.1, $\omega = \omega_0 \cup \omega_1$, ω contains at most isolated points, and has no accumulation points in U. So Lemma 1.4.2 is proved. $\qquad\square$.

Now we state a lemma for inner mapping (see definition below) (cf. [Kz, Ch.5]).

Lemma 1.4.3 Let $w = I(z)$ be continuous on domain D. Then the following three conditions are equivalent:

(i) Let $I(D_0)$ be the image of an open set D_0 in D. Then, $I(D_0)$ is open and the image of any continuum contains more than one point.

(ii) With the exception of at most some isolated points in D, the mapping is local homeomorphic.

(iii) There exists a homeomorphism $H : D \mapsto H(D)$ and an analytic function $A(\zeta)$ on $H(D)$ such that $I(z) = A(H(z))$.

If $I(z)$ satisfies one of the above conditions, then the mapping $I : D \mapsto I(D)$ is called an inner mapping.

Proof of Theorem 1.4.1.   From Lemma 1.4.2 and Lemma 1.4.3, we obtain (1.4.1). The function $f = f(\phi)$ is analytic on Ω ($\Omega = \phi(U)$). If $f^{(1)}(\phi) \equiv 0, \phi \in \Omega$, then $f(\phi) \equiv$ const. for $\phi \in \Omega$. Therefore $P(z) \equiv$ const. for all $z \in U$. This contradicts the assumption that $p(\zeta)$ is a spline function of degree $n, n \geq 1$ since $p \in \Phi_n(\Delta)$. Therefore, the set $\Omega_0 := \{\phi | f^{(1)}(\phi) = 0\}$ is discrete. Let ϕ_0 be a point in $\Omega \backslash \Omega_0$; then $f^{(1)}(\phi_0) \neq 0$. There are open sets $O(\phi_0)$ and $O(f_0), f_0 = f(\phi_0)$, such that the mapping $f : O(\phi_0) \mapsto O(f_0)$ is $1 - 1$, so the inverse function f^{-1} exists and

$$\phi(z) = f^{-1}(f(\phi(z))) = f^{-1}(p(z)) \quad \text{for} \quad z \in O(z_0)$$

because the analyicity of f^{-1} and the harmonicity of $P(z)$. Hence $\phi(z)$ is in $C^\infty(U)$. By simple calculation, we have

$$\bar{\partial}\partial\phi(z) = -\frac{f^{(2)}(\phi(z))}{(f^{(1)}(\phi(z)))^3}\bar{\partial}P(z) \cdot \partial P(z) \quad \text{for} \quad z \in O(z_0). \qquad (1.4.12)$$

From (1.4.12) we can prove the rest part of the theorem. In fact, since $\phi(z) = f^{-1}(P(z))$ for $z \in O(z_0)$,

$$\partial\phi(z) = (f^{-1})^{(1)}\partial P, \quad \bar{\partial}\partial\phi(z) = (f^{-1})^{(2)}\partial P \cdot \bar{\partial}P,$$

$$\frac{df^{-1}}{df}\frac{df}{d\phi} = 1, \quad \frac{d^2 f^{-1}}{df^2} = -\frac{f^{(2)}}{(f^{(1)})^3}.$$

If $\phi(z)$ is a harmonic function, then

$$f^{(2)}(\phi(z)) \cdot \bar{\partial}P(z) \cdot \partial P(z) = 0, \quad z \in O(z_0). \qquad (1.4.13)$$

We consider two cases:

(a) The zeros of the function $f^{(2)}(\phi)$ have a limit point in $O(\phi_0)$. We have $f^{(2)}(\phi) \equiv 0$ for $\phi \in \Omega$. Since $f^{(2)}(\phi)$ is an analytic function on $\Omega, f(\phi) = A\phi + B, A \neq 0$. Since otherwise $f(\phi)$ would be a constant, and

$P(z)$ a constant. This contradicts the hypothesis that the boundary value of $P(z)$ or $p(z)$ is a spline of degree $n, n \geq 1$. Thus $f(\phi) = A\phi + B, A \neq 0$.

(b) The zeros of the function $f^{(2)}(\phi)$ have no limit points in $O(\phi_0)$. If the set D_0 of zeros of the function $\partial P(z)$ is dense in $O(z_0)$, then $\partial P(z) = 0$ for all $z \in O(z_0)$ and hence $\partial P(z) = 0$ for $z \in U$. Thus $F_1(z) \equiv 0$ for $z \in U$ since $F_1(z)$ is analytic in U. From (1.4.10), we have $a_j = 0$ for $j = 0, 1, \cdots$ and $C_\nu = 0$ for $\nu = 1, 2, \cdots$. Thus

$$p(\zeta) = \sum_{\nu=-\infty}^{0} C_\nu \zeta^\nu, \quad \zeta = e^{i\theta}.$$

Since $p(\zeta)$ is a polynomial of degree n on each γ_j, we thus have $C_{-\nu} = 0, \nu = 1, 2, \cdots$, and $p(\zeta) = C_0$. This is impossible since $p(\zeta)$ is a spline of degree $n, n \geq 1$. Consequently, D_0 is not dense in $O(z_0)$. There is a point $\tilde{z}$ in $O(z_0)$ and a neighborhood $O(\tilde{z}) \subset O(z_0)$ such that $\partial P(z) \neq 0$ for $z \in O(\tilde{z})$ and hence $F_2(z) = 0$ for $z \in O(\tilde{z})$. Thereby, $F_2(z) = 0$ for $z \in U$. From the proof of Lemma 1.4.2, $p(\zeta)$ is a polynomial of degree n, and $P(z)$ is a polynomial of degree n. The converse is obvious. The proof of Theorem 1.4.1 is completed.　　　　　　　　　　　　　　　　　□.

To prove Theorem 1.4.2, we need some lemmas. Let $z = \lambda(t)$ be a continuous function defined on $[\alpha, \beta]$, $\lambda(t') \neq \lambda(t'')$ if $t' \neq t''$. Then, the function $z = \lambda(t)$ defines a curvilinear half interval half interval (Cu.H.I.). Let E_r be the set of all limit points on $z_n = \lambda(t_n)$ when t_n tends to β. If E_r contains only one point z', then we set $\lambda(\beta) = z'$. In this case, $\lambda(t)$ is a Jordan curvilinear half interval interval (J.Cu.I.) and its end point is z'.

The following lemmas can be found in [Mar, Ch.5, Sec.3].

Lemma 1.4.4　　If γ is a Cu.H.I. in the domain G, the limit points of γ lie on ∂G. Let $w = f(z)$ be a homeomorphism from G onto $\Delta = f(G)$. Then, $f(\gamma)$ is a Cu.H.I. with respect to the domain Δ, all its limit points lie on $\partial \Delta$.

Lemma 1.4.5　　Let $\gamma : z = \lambda(t), t \in [\alpha, \beta)$, be a Cu.H.I. in the unit disc U. The set E_r of all its limit points lies on ∂U and contains more than one point. If the function $f(z)$ is analytic and bounded on U and $\lim_{t \to \beta} f(\lambda(t)) = C$, then $f(z) \equiv C$ for all $z \in U$.

Let ∂G be the boundary of the domain G and $z' \in \partial G$ be the end point of a J.Cu.I. Then, z' is called an accessible point of ∂G. Such points are dense on ∂G. Two Cu.H.I. in G determine the same accessible point, if and only if the following two conditions are satisfied:

1. Γ_1, Γ_2 have a common end point ζ,

2. there is a Cu.H.I., Γ_3, such that in each neighborhood of ζ, Γ_3 intersects both Γ_1 and Γ_2.

Lemma 1.4.6　　Let G be a domain and U a disc. Suppose the mapping f maps G conformally onto U. Then f maps the different accessible points on ∂G onto different accessible points on ∂U.

Proof of Theorem 1.4.2.　　We shall prove that there is a decomposition of $P, P = \tilde{f} \circ \tilde{\phi}$, where $\tilde{\phi}$ is a homeomorphism from U onto $\tilde{\Omega}$, where $\tilde{\Omega}$ is a unit disc and $\tilde{f}$ is analytic on $\tilde{\Omega}$. By Theorem 1.4.1 $P = f \circ \phi$ and $\Omega = \phi(U)$. If Ω is the extended complex plane, then, by Liouvilles' theorem, $f(\phi) =$ const., hence $P(z) \equiv$ const. This contradicts the hypothesis of Theorem 1.4.2. If Ω is a punctuated plane, i.e., $\Omega = E \backslash \phi_0$, where E is the extended complex plane and ϕ_0 some points, then the function $f(\phi)$ has an isolated singular point at ϕ_0. Since $P(z)$ is bounded, ϕ_0 cannot be an essential singular point or a pole of $f(\phi)$, it must be a removable singular point. Thus we still have $P(z) \equiv$ const., a contradiction to the hypothesis of Theorem 1.4.2. We conclude that there are at least two points on $\partial\Omega$ and by Riemann's theorem there exists a conformal mapping $\varphi : \phi \mapsto Y$ which maps Ω onto a unit disc U_γ. Set $\tilde{f} = f \circ \varphi^{-1}, \tilde{\phi} = \varphi \circ \phi$. $\tilde{\phi}$ is a homeomorphism from U onto U_γ, and $\tilde{f}$ is analytic on U_γ. Thus $P(z) = \tilde{f}(\tilde{\varphi}(z))$. We will still write $P(z) = f(\varphi(z))$ for the sake of convenience, and $\Omega = \varphi(U)$ is a unit disc.

Suppose that there is a point $b \in \partial\Omega$ and b is the image of two accessible points $a_1, a_2, a_i \in \partial U, i = 1, 2$. Let

$$\tilde{\gamma}_i = \{\lambda_i(t), \alpha_i \leq t < \beta_i\}, \quad \lim_{t \to \beta_i} \tilde{\gamma}_i(t) = a_i, \quad \lim_{t \to \beta_i} \varphi(\gamma_i(t)) = b$$

, where $\tilde{\gamma}_i$ is a J.Cu.I., $i = 1, 2$. Let $C(b, \varepsilon)$ be a small disc with center b and radius ε. The set $O(b, \varepsilon) = C(a, \varepsilon) \cap \Omega$ is connected. Under the topological mapping $\varphi^{-1} : \Omega \mapsto U$ its image $O(a_1, a_2)$ is also a connected set, and $\overline{O}(a_1, a_2)$ is a continuum. Let ε tend to zero. Then $\overline{O}(b, \varepsilon)$ shrinks to the point b; meanwhile, $\overline{O}(a_1, a_2)$ shrinks to a continuum γ_{a_1,a_2} which contains a_1 and a_2 (for a more detailed proof, refer to the proof and the conclusion about its inverse φ below). Let z_0 be any point of $\gamma_{a_1 a_2}$. Then

$$\lim_{z \to z_0, z \in U} f(\varphi(z)) = f(b) = P(z_0),$$

thus $P(z) \equiv$ const., when $z \in \gamma_{a_1 a_2}$. Therefore, $P(z)$ must be a constant on an arc containing $\gamma_{a_1 a_2}$ since $P(z)$ is continuous on $\overline{U}$. But this contradicts the assumption of Theorem 1.4.2. Therefore we conclude that two accessible points cannot be mapped to one point. On the other hand, let γ_z be a J.Cu.I. in U. Its end point is b, i.e, $\gamma_z = \{z(t), \alpha \leq t < \beta\}, z(\beta) = b$. Let E_ϕ be the set of limit points $\{\phi(z(t_n))\}$ when t_n tends to β along γ_z. We assert that E_ϕ contains just one point. Otherwise, E_ϕ must be a continuum. If not, $E_\phi = E_1 \cup E_2, E_1 \cap E_2 = A$ (A : empty set), where E_1 and E_2 are closed sets, the distance between E_1 and E_2 is dis$(E_1, E_2) = 4\varepsilon, \varepsilon > 0$, and $O(E_i, \varepsilon) := \{z | \text{dis}(E_i, z) < \varepsilon\}, i = 1, 2$. Thus

$$O(E_1, \varepsilon) \cap O(E_2, \varepsilon) = A \quad \text{(empty set)} \tag{1.4.14}$$

There exist two sequences $\{t_n\}$ and $\{\tau_n\}$ of points such that $\lim_{n \to \infty} t_n = \lim_{n \to \infty} \tau_n = \beta$. Moreover,

$$\lim_{n \to \infty} z(t_n) = \lim_{n \to \infty} z(\tau_n) = b. \tag{1.4.15}$$

There also exist integers n_A and n_B such that if $n \geq \max(n_A, n_B)$, then

$$\phi(z(t_n)) \in O(E_1, \varepsilon), \quad \phi(z(\tau_n)) \in O(E_2, \varepsilon). \tag{1.4.16}$$

From the fact that $t_n \to \beta$ and $\tau_n \to \beta$, we assert that there exist two subsequences $\{t_{n_j}\}$ and $\{\tau_{m_k}\}$ such that

$$t_{n_1} < \tau_{n_1} < t_{n_2} < \tau_{n_2} < \cdots \qquad (1.4.17)$$

where $t_{n_j} \to \beta$ as $j \to \beta$, and $\tau_{m_k} \to \beta$ as $k \to \infty$.

From (1.4.14) and the connectivity of $\gamma_\varphi = \varphi(\gamma_z)$, there is a point ξ_j in the interval (t_n, τ_{m_j}), such that

$$\phi(z(\xi_j)) \notin \{O(E_1, \varepsilon) \cup O(E_2, \varepsilon)\}. \qquad (1.4.18)$$

From (1.4.17), $\lim_{j \to \infty} \xi_j = \beta$, and the set $\tilde{E}$, of the limit points, of $\{\phi(z(\xi_j))\}$ is in E_ϕ. On the other hand, by (1.4.18), $\tilde{E}$ is neither in $O(E_1, \varepsilon)$ nor in $O(E_2, \varepsilon)$. Therefore, $\tilde{E}$ does not belong to $E_\phi = E_1 \cup E_2$. This is a contradiction. Hence E_ϕ is a continuum, the accessible points are dense in E_ϕ. Since $\gamma_z = \{z(t), \alpha \le t < \beta\}$ is a J.Cu.I. in U, from Lemma 1.4.4, $\phi(z)$ maps γ_z onto $\gamma_\phi = \{\phi = \phi(z(t), \alpha \le t < \beta\}$ and γ_ϕ is a Cu.J.I. in Ω. f maps γ_ϕ onto γ_f. By Lemma 1.4.5, $f(\phi) = $ const. for $\phi \in \Omega$, therefore, $P(z) = $const., $z \in U$, which contradicts the assumption of Theorem 1.4.2. Thus E_ϕ contains only one point. We have then proved that there is an extension of the homeomorphism ϕ from U to $\overline{U}$, which maps $\overline{U}$ topologically onto $\overline{\Omega}$.   □.

Now we prove Theorem 1.4.4.

Since $P(z) = f(\phi(z)), z \in U$, from the hypothesis of Theorem 1.4.4, the set of zeros of the analytic function f has no accumulation points in Ω. Since $P(\zeta) \ne 0$, for $\zeta \in \Gamma$, by assumption, we conclude that $f(\phi) \ne 0, \phi \in \partial\Omega$. Therefore, f has a finite number of zeros in $\overline{\Omega}$. So we can write

$$P(z) = f(w(z)) = \prod_{j=1}^{N} (\phi(z) - \phi(z_j)) \hat{f}(\phi(z)),$$

where $\hat{f}$ is analytic on Ω and has no zeros on $\overline{\Omega}$. Because $P(z)$ and $\phi(z)$ are continuous on $\overline{U}$, $f(\phi)$ must be continuous on $\overline{\Omega}$. We use the argument principle of analytic functions,

$$N = \frac{1}{2\pi} \triangle_{\partial\Omega} \arg f(\phi) = \frac{1}{2\pi} \triangle_\gamma \arg P(z).$$

Since

$$\Gamma = \bigcup_{j=1}^{K} \gamma_j, \quad \gamma_j = \gamma(z_j, z_{j+1}),$$

set $p(\zeta) = p_j(\zeta)$, $\zeta \in \gamma_j$, and $p_j(\zeta)$ is a polynomial of degree n.

$$
\begin{aligned}
N &= \frac{1}{2\pi} \Delta_\Gamma \arg P(\zeta) = \frac{1}{2\pi} \sum_{j=1}^{K} \Delta_{\gamma_j} \arg p_j(\zeta) \\
&< \frac{1}{2\pi} \sum_{j=1}^{K} \Delta_\Gamma \arg p_j(\zeta) \\
&= \frac{K}{2\pi}(2n\pi) = nK.
\end{aligned}
$$

This completes the proof of (1.4.2).   □.

We have explored the nature of CHSF in details. By the maximum modulus principle, if an analytic function F is continuous on $\overline{U}$, F can then be approximated on $\overline{U}$ by a CHSF if the boundary values of both functions are close. In particular, if the boundary function of a conformal mapping can be approximated by a CHSF to any given accuracy, so can the conformal mapping function on $\overline{U}$.

§1.5   Applications of CHSF to Approximation of conformal mappings

Let γ be a piecewise analytic Jordan curve, D the interior domain enclosed by γ and F a conformal mapping from D onto unit disc U. By the Osgood–Caratheodory theorem (cf.[H]), F can be extended to a topological map of the closure of D onto the closure of U. Let the parametric representation of γ be given by

$$z = z(\tau), \quad 0 \le \tau \le \beta \tag{1.5.1}$$

As $z(\tau)$ moves along γ, the image point $F(z(\tau))$ describes the unit circle $\Gamma(= \partial U)$. Thus

$$\theta(\tau) := \arg F(z(\tau)), \quad 0 \le \tau \le \beta,$$

may be defined as a continuous function. Such continuous argument of $F(z(\tau))$ is called an interior boundary correspondence function for the map F.

The following theorem is important in the construction theory of conformal mappings (cf. [H]):

Theorem 1.5.1 Let γ be a piecewise analytic Jordan curve and $\theta(\tau)$ be an interior boundary correspondence function of γ. Then the following equation

$$\frac{1}{2\pi} \int_0^\beta \ln|z(\sigma) - z(\tau)|\mu(\tau)d\tau = \ln|z(\sigma)|, \quad 0 \le \sigma \le \beta, \qquad (1.5.2)$$

has a solution $\mu(\tau) = \theta'(\tau)$ in every space $L_p(0,\beta)$ where $p < 2$. If the interior angles at all corners are less than $< 2\pi$, then this solution is in $L_2(0,\beta)$.

The solution of (1.5.2) is unique if the function $\mu(t)$ satisfies

$$\int_0^\beta \mu(t)dt = 2\pi. \qquad (1.5.3)$$

Then the function $\theta(\tau) := \int_0^\tau \mu(t)dt$ satisfies the following conditions:

$$\theta(0) = 0, \quad \theta(\beta) = 2\pi. \qquad (1.5.4)$$

using numerical method, we have

$$\theta(\tau) = \sum_{j=0}^m \alpha_j e_j(\tau), \quad 0 \le \tau \le \beta \qquad (1.5.5)$$

where $\{e_j\}_{j=0}^\infty$ is a basis in $L_2(0,\beta)$.

If we choose $\{e_j\}_0^\infty$ to be the set of functions $\{1, t, \cos kt, \sin kt\}_{k=1}^\infty$, then (1.5.5) has the following form

$$\theta(\tau) = \tau + \sum_{j=1}^n \frac{\alpha_j}{j} \sin j\tau - \sum_{j=1}^{n-1} \frac{\beta_j}{j} \cos j\tau + \sum_{j=1}^{n-1} \frac{\beta_j}{j} \qquad (1.5.6)$$

$0 \le \tau < 2\pi$, where α_j, β_j are constant, and assume that $\beta = 2\pi$.

By solving the system of equations:

$$\theta(\tau) = \frac{2\pi l}{K}, \quad l = 1, \cdots, K \tag{1.5.7}$$

the data of the boundary function $F : \{F_l\}_1^K = \{F(e^{i2\pi l/K})\}_1^K$ are obtained (cf. [CH1], [CH2], [R1], [R2]).

Now $\theta(\tau)$ is solved, $z(\tau)$ is the parametric representation of γ, then the mapping function F can be constructed as follows:

$$F(z) = z \exp\left\{-\frac{1}{2\pi}\int_0^{2\pi} \ln\left[1 - \frac{z}{z(\tau)}\right]\theta^{(1)}(\tau)d\tau\right\}, \quad z \in D$$

$$F'(0) = \exp\left\{-\frac{1}{2\pi}\int_0^{2\pi} \ln|z(\tau)|\theta^{(1)}(\tau)d\tau\right\} \tag{1.5.8}$$

(cf. [H, p.383]).

The conformal mapping from U onto D is more difficult to construct. In the following we shall give an approximate mapping by using CHSF.

We have proved in Theorem 1.4.3 that the CHSF defined in Theorem 1.4.1 can be decomposed into $P = f(\phi)$, from Theorem 1.4.3, if $\gamma = P(\Gamma)$ is a Jordan curve, then the mapping $P, U \mapsto P(U)$ is a one to one mapping, and f is a conformal mapping.

There are several approaches to construct the boundary function g of the CHSF

$$P(w) = \frac{1}{2\pi}\int_0^{2\pi} g(\zeta)\mathrm{Re}\left(\frac{\zeta + w}{\zeta - w}\right)d\theta, \quad \zeta = e^{i\theta}, w \in \overline{U}.$$

We list them below:

(1) the interpolating function from $\Phi_n(\Delta)$,
(2) the quasi-interpolant from $\Phi_n(\Delta)$,
(3) the projection method.

Theorem 1.5.2   Let $F(w)$ be a complex harmonic function in U, continuous on $\overline{U}$. $P(w)$ is the CHSF defined by

$$P(w) = \frac{1}{2\pi}\int_0^{2\pi} S(e^{i\theta})\mathrm{Re}\left(\frac{e^{i\theta} + w}{e^{i\theta} - w}\right)d\theta, \quad w \in \overline{U}, \tag{1.5.9}$$

where $S(e^{i\theta})$ is a complex cubic spline satisfies

$$S(z_j) = f_j, \quad j = 1, \cdots, K \tag{1.5.10}$$

where $\{f_j\}_1^K$, $f_j = F(z_j)$, are given data. Then, $P(w)$ converges uniformly to $F(w)$ in the closed disc $\overline{U}$ as $|\triangle| \to 0$. Moreover, the following estimates are valid

$$|F(w) - P(w)| < L(R)\, \omega(F, |\triangle|), \quad w \in \overline{U} \tag{1.5.11}$$

where

$$\begin{aligned}
L(R) &= \min(5.13R + 7.13, 0.07R^2 + 1.5), \\[2mm]
R &= \frac{\max_j |z_{j+1} - z_j|}{\min_j |z_{j+1} - z_j|}
\end{aligned} \tag{1.5.12}$$

ω is the modulus of continuity of F on Γ, and

$$|\triangle| = \max_j |z_{j+1} - z_j|.$$

Proof.   Let the error function $E(w)$ be $F(w) - P(w)$. By Schwartz's theorem (cf. [T, p130, Th.IV,2]), $E(w)$ is continuous on $\overline{U}$, hence $E \in H^{(0)}(\overline{U})$ (§1.3). By Theorem 1.1.4 and maximum modulus principle, (1.5.11) follows immediately. $\square$.

The following theorem indicates that the degree of approximation increases if the curve γ becomes smoother.

Theorem 1.5.3   Let γ satisfy the Ljapunov condition. Let $z = F(w)$ be the mapping function from U onto D, $S(w)$ the complex cubic spline which interpolates $F(w)$ at $\triangle = \{z_j\}_1^K$, and $P(w)$ the CHSF with boundary function $S(w)$. Then, for all $w \in \overline{U}$,

$$|P(w) - F(w)| \le M_1 |\triangle|^{1+\alpha}, \tag{1.5.13}$$

$$|P_w(w) - F'(w)| \le \xi |\triangle|^{\alpha/2}, \tag{1.5.14}$$

$$|P_{\overline{w}}(w)| \le \xi |\triangle|^{\alpha/2}, \tag{1.5.15}$$

where

$$\xi = (2^{\alpha/2}/\alpha)M_3 + M_2|\triangle|^{\alpha/2},$$

$$M_1 = \frac{\pi}{2}M_2, \quad M_2 = (26 + 14\sqrt{2})\frac{M}{\alpha+1}\left(\frac{\pi}{2}\right)^{\alpha+1},$$

$$M_3 = 2\left(\frac{38 + 40\sqrt{2}}{\alpha+1}\cdot\frac{\pi}{2} + 1\right)\left(\frac{\pi}{2}\right)^{\alpha} MR,$$

$$R = \frac{\max_j |z_{j+1} - z_j|}{\min_j |z_{j+1} - z_j|},$$

M, α are Ljapunov constants,

$$|F'(e^{i\theta_1}) - F'(e^{i\theta_2})| \leq |\theta_1 - \theta_2|^{\alpha},$$

and

$$P_w = \partial_w P; \quad P_{\overline{w}} = \partial_{\overline{w}}P.$$

Proof. (1.5.13) is obtained directly from the maximum modulus principle and (1.1.38). Now we prove (1.5.14). Set

$$\varphi(\zeta, w_0) = (S'(\zeta) - F'(\zeta)) - (S'(w_0) - F'(w_0)). \qquad (1.5.16)$$

From (1.1.40)

$$|\varphi(\zeta, w_0)| \leq M_3|\triangle|^{\alpha-\delta}|\zeta - w_0|.$$

Since $\displaystyle\int_{\Gamma} |\zeta - w_0|^{\delta-1}|d\zeta| \leq 2^{\delta}\pi/\delta$, set $\delta = \frac{\alpha}{2}$, then $\qquad (1.5.17)$

$$\left|\frac{1}{2\pi}\int_{\Gamma} \frac{\varphi(\zeta, w_0)}{\zeta - w_0}d\zeta\right| \leq \frac{2^{\alpha/2}}{\alpha}M_3|\triangle|^{\alpha/2}. \qquad (1.5.18)$$

From (1.5.16), (1.5.18) and (1.1.39), we have

$$|P_w(w) - F'(w)|$$

$$\leq \sup_{w_0\in\Gamma}\left\{\left|\frac{1}{2\pi}\int_{\Gamma} \frac{\varphi(\zeta, w_0)}{\zeta - w_0}d\zeta + S'(w_0) - F'(w_0)\right|\right\} \qquad (1.5.19)$$

$$\leq \left[\frac{2^{\alpha/2}}{\alpha}M_3 + M_2|\triangle|^{\alpha/2}\right]|\triangle|^{\alpha/2} = \xi|\triangle|^{\alpha/2}.$$

Then (1.5.14) is proved.

From (1.1.40), (1.1.39) and (1.5.17), for $w \in \overline{U}$

$$|P_{\overline{w}}(w)| \leq \frac{1}{2\pi} \sup_{w_0 \in \Gamma} \left| \int_\Gamma \frac{(\zeta S'(\zeta) - w_0 S'(w_0)) - (\zeta F'(\zeta) - w_0 F'(w_0))}{\zeta - w_0} d\zeta \right|$$
$$\leq \frac{1}{2\pi} \{ M_3 |\triangle|^{\alpha-\delta} 2^\delta \delta^{-1} + 2\pi M_2 |\triangle|^\alpha \}.$$

Set $\delta = \alpha/2$ we then have (1.5.15) (cf. [CH1]). $\square$.

From Theorem 1.5.2 and Theorem 1.5.3, CHSF $P(w)$ is indeed a good approximation to the conformal mapping F since both the CHSF $P(w)$ and its derivative $P_w(w)$ converge uniformly to $F(w)$ and $F'(w)$ on $\overline{U}$ respectively, if $|\triangle| \to 0$.

We can also use the quasi-interpolant as the boundary function of CHSF if the derivatives of the approximated function are easily calculated on the boundary of the domain. As for the projection method, we shall discuss it after when we study the construction of the periodic quasi-wavelets.

§1.6   Algorithm for Computing $P(z)$. (See [CH1] and [CH2])

In §1.3 we have decomposed the CHSF $P(z)$ into elementary functions as stated in Theorem 1.3.1. All terms in formulas given in (1.3.5), except $\varphi_{j,j+1}$ and $\psi_{j,j+1}$, are easily calculated. In this section we shall give the algorithms for the angles.

Recall the definitions of $\varphi_{j,j+1}(z)$ and $\psi_{j,j+1}(\hat{z})$ are given explicitly in the beginning of §1.3, from (1.3.2). We set

$$\lambda = \cos^{-1}\{[(x_j - x)(x_{j+1} - x) + (y_j - y)(y_{j+1} - y)] \qquad (1.6.1)$$
$$\cdot [(x_j - x)^2 + (y_j - y)^2]^{-\frac{1}{2}} [(x_{j+1} - x)^2 + (y_{j+1} - y)^2]^{-\frac{1}{2}}\},$$

where $z = x + iy, z_k = x_k + iy_k$.

Let C_2 be the complex plane, and $l_{j,j+1}$ be the straight line joining the points z_j and z_{j+1}. C_2 is divided by $l_{j,j+1}$ into Ω_1 and Ω_2, say, the half plane Ω_2 contains the circular arc γ_j (we stipulate that the length of

$\gamma_j, |\gamma_j| < \pi$). Then $\varphi_{j,j+1}$ and $\psi_{j,j+1}$ can be represented as follows:

$$\varphi_{j,j+1}(z) = \int_{z_j}^{z_{j+1}} d \arg(\zeta - z) = \begin{cases} \lambda, & z \in U \cap \Omega_1, \\ \pi, & z \in U \cap l_{j,j+1}, \\ 2\pi - \lambda, & z \in U \cap \Omega_2, \end{cases} \qquad (1.6.2)$$

where λ is defined as in (1.6.1) and

$$\psi_{j,j+1}(\hat{z}) = \int_{z_j}^{z_{j+1}} d \arg(\zeta - \hat{z}) = \begin{cases} -\lambda, & \hat{z} \in CU \cap \Omega_2, \\ 0, & \hat{z} \in CU \cap l_{j,j+1}, \\ \lambda, & \hat{z} \in CU \cap \Omega_1, \end{cases} \qquad (1.6.3)$$

where λ is defined as before, $\hat{z} = (\bar{z})^{-1} = \hat{x} + i\hat{y}$.

The following criterion tells us whether a point $\xi \in C_2$ lying in $\Omega_1(\Omega_2)$ or not.

Criterion

$$\xi \in \Omega_1, \qquad \text{iff} \quad |\xi| = ||\xi - z_0| - |z_0||, \qquad (1.6.4)$$

$$\xi \in \Omega_2, \qquad \text{iff} \quad |\xi| = |x_0| + |\xi - z_0|, \qquad (1.6.5)$$

$$\xi \in l_{j,j+1}, \qquad \text{iff} \quad |\xi| = |x_0|, \qquad (1.6.6)$$

where z_0 is the intersection of the straight line l with $l_{j,j+1}$; l is the line joining the point ξ and the origin $z = 0$.

In the programming, instead of (1.6.4)–(1.6.6), we use

$$|\xi| > ||\xi - z_0| - |z_0|| - \varepsilon, \qquad (1.6.7)$$

$$|\xi| < |z_0| + |\xi - z_0| + \varepsilon, \qquad (1.6.8)$$

$$|x_0| + \varepsilon \geq |\xi| \geq |z_0| - \varepsilon, \qquad (1.6.9)$$

respectively, ε is a small number such that

$$|\varepsilon| \sim 10^{-8}.$$

By applying above algorithms we plot graphs of different complex harmonic spline functions (cf. [CH2]), the procedure is as follows.

The input of the Jordan curve $\gamma = \{z|z = z(\tau), 0 \leq \tau \leq \beta\}$ yields the output of the picture of CHSF $P(w)$.

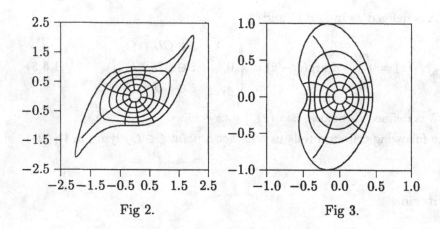

Fig 2.   Fig 3.

In Fig 2 the input Jordan curve is:

$$w(t) = 2\cos t + i(\sin t + 2\cos^3 t), \qquad 0 \leq t \leq 2\pi$$

In Fig 3 the input Jordan curve is:

$$\begin{aligned} w(t) &= \lambda(t) + i\omega(t) \\ \lambda(t) &= 2.25(0.2\cos t + 0.1\cos 2t - 0.1) \\ \omega(t) &= 2.25(0.35\sin t + 0.14\sin 2t - 0.02\sin 4t) \\ & \quad 0 \leq t \leq 2\pi \end{aligned}$$

In both figures, we choose 50 uniform knots on the unit circle for our interpolation. We plot the conform mapping of six circles: $z = re^{it}$, $0 \leq t \leq 2\pi$, where r equals to $\frac{1}{6}$, $\frac{2}{6}$, $\frac{3}{6}$, $\frac{4}{6}$, $\frac{5}{6}$ and 1 respectively. We also plot the mapping of the lines: $z = re^{it}$, $\frac{1}{7} \leq r \leq \frac{6}{7}$, where $t = j2\pi/10$, j is a integer which ranges from 1 to 10 respectively.

We can see that at the corresponding intersection points, the images of the orthogonal curves are 'almost' orthogonal in the sense that we cannot see that they are not.

§1.7   The Mappings between two Arbitrary Domains

Assume we are given two simply connected domains D' and D.

If we want to construct an approximant $F(D', D, z)$ for the conformal mapping G which maps D' onto D, then $F(D', D, z)$ is just a composition of two mapping functions $F(z)$ and $P(\zeta)$:

$$F(D', D, z) = P(F(z))$$

where $F(z)(z \in D')$ is the function in (1.5.8), it maps D' onto the unit disc U and $P(\zeta)(\zeta \in U)$ maps U onto D by formula (1.3.1). $F(D', D, z)$ is thus a complex harmonic function, it approximates the conformal mapping G with high accuracy if the length of step of the boundary spline function is small enough.

Formula (1.5.8) is an explicit expression. $P(\zeta)$ is a CHSF, and it can be decomposed into elementary functions. So we can use the algorithm introduced in the previous section.

In particular, if D' is a unit disc then we choose $F(z)$ to be the identity mapping, and $F(D', D, z) = P(z)$. If D is a unit disc, then we choose P to be the identity mapping, and then $F(D', D, z) = F(z)$. In general, D' and D are two arbitrary Jordan domains.

Note 1.   The development of conformal mapping has undergone a very long history, but still full of vigour and vitality. The main reason is that it unearths many new applications in various areas of applied sciences, such as ion optics, propagation of optical modes in dielectric fibers, diffraction of electromagnetic waves, atomic physics, radio propagation in the atmosphere, nonlinear diffusion problems, supersonic flows, etc. (cf.[SL]). In this book we first exhibit an approximation method provided by complex harmonic spline function (CHSF), which is very efficient and has high accuracy. But before we use the CHSF we have to know its boundary values.

In this chapter, by solving Symm's equation we obtain the discrete boundary values. In chapter 3 we shall introduce another method appealing to so-called quasi-wavelets to solve a kind of integral equations. Meanwhile, the projection methods are used, and it provides computational complexity of order $O(N)$.

Note 2.   Since one of the applications of CHSF is to approximate the conformal mappings, from §1.3 and §1.4 we see that one could not keep the specific properties as described in §1.3 and §1.4 if we use a real spline function as the boundary function of CHSF. Indeed, since the graph of a real spline function cannot be a contour, therefore we have to use the complex spline function as the boundary function of $P(z)$.

Note 3.   The interpolatory properties of complex splines on uniform meshes have been investigated in several papers ([ANW1], [MS], [SCh1], [SC], [Tz]). As for the interpolation of cubic complex splines on non-uniform meshes, Ahlberg, Nilson and Walsh gave some primary results [ANW2]. In [Xu] Xu Shi-ying pointed out that their results would be false if no additional condition were imposed. His point is as follows: let K be a smooth closed curve in the complex plane, and the set of points $\{t_j\}_1^N, t_{N+1} = t_1$ be scattered anti-clockwise on K, $\Delta : t_1 \prec t_2 \prec \cdots \prec t_j \prec t_{j+1} \prec t_N \prec t_{N+1} = t_1$. K_j denotes the subarc of K from t_{j-1} to t_j, s_j denotes the arc length from t_1 to t_j. Set $h_j = t_j - t_{j-1}, \overline{\Delta} = \max_j |h_j|, \overline{\overline{\Delta}} = \max_j(s_j - s_{j-1})$. In [ANW2], Ahlberg, et al., attempted to prove the existence and uniqueness of the interpolatory complex cubic spline on K. Xu Shi-ying in his paper [Xu] observed that the additional condition is that $\overline{\overline{\Delta}}$ must be small. Otherwise, the results in [ANW2] would be false. In 1981 [C3] Chen Han-lin proved the existence and uniqueness of the interpolatory complex splines (of degree 2 and 3), where the knots are arbitrarily spaced on the unit circle and no additional conditions are prescribed. These results are presented in §1 Theorem 1.1.1 and Theorem 1.1.2. The proofs of these theorems are quite different from others (see [ANW1] and [ANW2]). After that, Chen Han-lin and T. Hvaring gave the error estimates stated in Theorem 1.1.4 and Theorem 1.1.5.

In 1984 Gunther Schmidt proved the existence of complex spline of arbitrary odd degree with arbitrarily spaced knots on the unit circle [Schm]. Chen Tian-ping [Ct] generalized Schmidt's results to a complex spline of odd degree with arbitrary deficiency.

It is well known that one can count the number of zeros of the real-valued splines by the fundamental theorem of algebra. In 1983, Chen Han-lin gave a sharp upper bound for the number of the zeros of a certain class of complex splines [C5, Th.3].

Counting the number of zeros of circle splines (complex spline function on the unit circle) can be reduced to the relevant real spline functions as was done in [C3], and the interpolation for wide classes of functions can be solved by using Theorem 1 given in [C6]. Although the mixed boundary conditions imply many cases, yet the case of circle splines is excluded.

Based on the work of [GL2], Goodman and Lee gave a complete result for the interpolation of complex splines on the unit circle with arbitrarily spaced knots [GL1]; it is introduced in Theorem 1.1.3, and we omit its proof here.

Other kinds of complex spline functions, such as complex Λ-splines and complex discrete splines, have also been studied. The interested reader can refer to [MS], [GLS], [W] and [MaS]. The complex spline functions on the plane were studied by Chen Han-lin [C7] and G. Opfer [W].

A more useful type of complex spline on the circle is the complex quasi-interpolant. This can be found in [C1] and [GLS], [GL1].

Note 4. As we know, the analytic spline function was first presented in Ahlberg's paper [Ah], in 1969. The analytic splines converge to the approximated function only in the interior of the unit disc. More precisely, these approximants only converge uniformly on a compact subset of the open disc. Z. Wronicz constructed a basis $\{G_n^{(m,k)}\}$ of analytic functions with the aid of the Schwartz formula [Wr1], [Wr2]. The construction procedure of $G_n^{(m,k)}$ is rather involved and the boundary value of the Schwartz integral is much more complicate than that of CHSF.

From the applied point of view, for choosing the space of functions of approximation it is important to emphasize the following three points: (i) simplicity for the formulation; (ii) each error of different order converges uniformly to zero on the closed disc $\overline{U}$ if the sizes of meshes go to zero; (iii) high accuracy — the error of each order of the derivatives of the approximated function by the approximating function can be small when the size of the mesh is small. If we choose the CHSF as the approximating functions then the accuracy (i.e., (iii)) is guaranteed by Theorem 1.3.2 and Corollary 1.3.1. Since the CHSF can be decomposed into elementary functions, as described in (1.3.4) and (1.3.5), so the first point (i) is also retained. The second point is also true since the estimates in (1.3.12), (1.3.13), (1.3.14) and the inequality in Corollary 1.3.1 are valid on $\overline{U}$.

Chapter 2

PERIODIC QUASI-WAVELETS

§2.1 Periodic Orthonormal Quasi-Wavelets

In this section we introduce the so-called periodic orthonormal quasi-wavelets. The kind of wavelet which we want to construct possesses orthonormality; the numbers of terms in the decomposition and reconstruction formulas are strictly limited, the localization is not emphasized, and such a kind of wavelet we call quasi-wavelets.

Later we shall see these classes of quasi-wavelets play a key role in solving some integral equations. Combining the new techniques of multi-strategy, we can greatly reduce the computational complexity in the computational procedure (cf. Chapter 3).

We begin with the spline functions. Those who are familiar with the definition and the basic properties of spline functions may skip over this part.

Let $\mathbb{Z}$ be the set of all integers, $\mathbb{Z}_+$ denote the set of all positive integers, $\mathbb{N}$ be the set of non-negative integers.

Let K, n be in $\mathbb{Z}_+$ and $\mathbb{N}$ respectively, $K \geq n + 2$, h a real number, and $T = Kh$. Set $n_0 = (n+1)/2, y_0 = -n_0 h, y_j = y_0 + jh, j = 1, 2, \cdots$. The B-spline (Basic spline) is defined as follows

Definition 2.1.1 Let $[y_i, \cdots, y_{i+n+1}]_y f(y)$ be the divided difference over the points $y_i, \cdots, y_{i+n+1}$, i.e.,

$$[y_i, \cdots, y_{i+n+1}]_y f(y) = \sum_{j=i}^{i+n+1} f(y_j)/\omega'(y_j) \qquad (2.1.1)$$

where $\omega(y) = (y - y_i) \cdots (y - y_{i+n+1})$. The B-spline $B_i^n(x)$ can be defined as follows:

$$B_i^n(x) = (-1)^{n+1}(y_{i+n+1} - y_i)[y_i, \cdots, y_{i+n+1}]_y(x - y)_+^n \qquad (2.1.2)$$

57

where

$$(x - y)_+^n := \begin{cases} (x - y)^n & \text{if } x \geq y \\ 0 & \text{otherwise} \end{cases} \qquad (2.1.3)$$

$n \in \mathbb{N}$.

We have the following properties of B-spline functions:

[P1]   from (2.1.2) and (2.1.1), $B_i^n(x)$ can also be written as

$$B_i^n(x) = \frac{1}{n! h^n} \sum_{j=0}^{n+1} (-1)^j \binom{n+1}{j} (x - y_{i+j})_+^n; \qquad (2.1.4)$$

[P2] $\text{supp} B_i^n = [y_i, y_{i+n+1}] = [(i - n_0)h, (i + n_0)h] \qquad (2.1.5)$

$$= \left[\left(i - \frac{n+1}{2}\right) h, \left(i + \frac{n+1}{2}\right) h \right];$$

[P3] $B_i^n(x) > 0, x \in (y_i, y_{i+n+1}); \qquad (2.1.6)$

[P4] $B_i^n \in C^{n-1}(R^1)$, and the knots of B_i^n are $\{y_\nu\}_i^{i+n+1}$;

[P5] $B_{i+\nu}^n(x) = B_i^n(x - \nu h), \quad \nu \in \mathbb{Z} \qquad (2.1.7)$

here we have extended the sequence $y_i = y_0 + ih$ of points, where i may be any integer.

[P6] $D_x^k B_i^n(x) = \frac{1}{h^k} \sum_{\nu=0}^{k} (-1)^\nu \binom{k}{\nu} B_i^{n-k}\left(x + \frac{k - 2\nu}{2} h\right),$

for $k = 0, \cdots, n$. $D_x^k = \dfrac{d^k}{dx^k}$ and

$$D_x^{n+1} B_i^n(x) = \frac{1}{h^n} \sum_{\nu=0}^{n+1} (-1)^\nu \binom{n+1}{\nu} \delta\left(x - \left(i + \nu - \frac{n+1}{2}\right) h\right);$$

$$(2.1.8)$$

[P7] $\sum_{j \in \mathbb{Z}} B_j^n(x) = 1, \quad \forall x \in R^1. \qquad (2.1.9)$

Naturally, we can use the knots $\{y_\nu^m\}_{\nu \in \mathbb{Z}}$ instead of $\{y_\nu\}_{\nu \in \mathbb{Z}}$, where

$$y_\nu^m = y_0^m + \nu h_m,$$
$$y_0^m = -n_0 h_m, \quad h_m = T/K(m), \quad K(m) = 2^m K, \quad m \in \mathbb{N}. \qquad (2.1.10)$$

Evidently, $y_\nu^0 = y_\nu$. With the knots $\{y_\nu^m\}$, we also can define B-spline $B_j^{n,m}(x)$ or $B_j^n(x, h_m)$ as in (2.1.2):

$$B_j^{n,m}(x) := B_j^n(x, h_m)$$
$$:= (-1)^{n+1}(y_{i+n+1}^m - y_i^m)[y_i^m, \cdots, y_{i+n+1}^m]_y (x - y)_+^n. \qquad (2.1.11)$$

$B_j^{n,m}(x)$ enjoys all properties listed above in [P1]–[P7], but now the step length is h_m instead of h in all formulas.

The family of all such splines with knots $\{y_\nu^m\}_{\nu\in\mathbb{Z}}$ will be denoted by $S_n(h_m)$, where n is the degree of splines and h_m the length of step.

Now we define the class of periodic spline functions in $S_n(h_m)$.

$$\tilde{S}_n(h_m): \ = \tilde{S}_n(h_m, [0, T])$$
$$= \{f \mid f \text{ is a polynomial of degree } n \text{ on each interval}$$
$$[y_j^m, y_j^m + h_m), j = 0, 1, \cdots, K(m) - 1;$$
$$f \in C^{n-1}[0, T] \text{and} S^{(i)}(0) = S^{(i)}(T),$$
$$i = 0, \cdots, n - 1\}.$$

We construct the basis for $\tilde{S}_n(h_m)$.

Theorem 2.1.1   Set the function $\tilde{B}_i^{n,m}(x)$:

$$\tilde{B}_i^{n,m}(x) = B_i^{n,m}(x) + B_{i+K(m)}^{n,m}(x) \tag{2.1.12}$$

and the integer $\hat{n}_0 := \left[\frac{n}{2}\right] + 1$, then the system of functions

$$\Omega_n^m := \{\tilde{B}_i^{n,m} \mid i = -\hat{n}_0, \cdots, K(m) - \hat{n}_0 - 1\}$$

forms a basis for $\tilde{S}_n(h_m)$.

Proof.   For n odd or even, each function $S(x)$ in $\tilde{S}_n(h_m)$ can be presented as follows

$$\begin{cases} S(x) = \sum_{j=0}^{n} a_j x^j + \sum_{\nu=\hat{n}_0+1}^{\hat{n}_0+K(m)-1} b_j (x - y_j^m)_+^n, & x \in [0, T] \\ S^{(j)}(0) = S^{(j)}(T), & j = 0, \cdots, n - 1 \end{cases} \tag{2.1.13}$$

where a_j, b_j are constants.

There are $n + K(m)$ coefficients in (2.1.13), but subject to n constrains, the degree of freedom is $K(m)$. Since $\tilde{B}_i^{n,m}$ defined in (2.1.12) belongs to $\tilde{S}_n(h_m)$, the supports of any two functions in Ω_n^m are different [P2], therefore the functions in Ω_n^m are linearly independent. Thus the dimension of $\tilde{S}_n(h_m)$ is $K(m)$ and Ω_n^m is a basis of $\tilde{S}_n(h_m)$.   □.

Each function in $\tilde{S}_n(h_m)$ can be extended periodically to the whole real axis. The collection of the extended periodic functions in $\tilde{S}_n(h_m)$ is denoted by $\overset{\circ}{S}_n(h_m)$. Evidently, $\tilde{S}_n(h_m)$ is the restriction of $\overset{\circ}{S}_n(h_m)$ on the interval $[0,T]$.

In the following we give the Fourier expansion of $\overset{\circ}{B}_0^{n,m}(x)$— the periodic extension of $\tilde{B}_0^{n,m}(x)$.

Theorem 2.1.2   The Fourier expansion of $\overset{\circ}{B}_0^{n,m}(x)$ is

$$\overset{\circ}{B}_0^{n,m}(x) = (K(m))^n \sum_{\gamma \in \mathbf{Z}} \left(\frac{\sin \dfrac{\nu\pi}{K(m)}}{\nu\pi} \right)^{n+1} \exp\left(\frac{i2\pi\nu x}{T} \right), \quad x \in R^1.$$

$$(2.1.14)$$

Proof.   Define a function $\overset{\circ}{M}_0^{n,m}(x)$ as follows:

$$\overset{\circ}{M}_0^{n+1,m}(x) = \frac{1}{T} \int_0^T \overset{\circ}{M}_0^{n,m}(x-y) \overset{\circ}{M}_0^{0,m}(y) dy, \quad n \in \mathbf{N}, \qquad (2.1.15)$$

where $\overset{\circ}{M}_0^{\nu,m}(x)$ is a periodic extension of $M_0^{\nu,m}(x)$. $M_0^{0,m}(x) := \tilde{B}_0^{0,m}(x)$, then $M_0^{0,m}(x) = 1$ if $x \in I_m := \left[0, \frac{1}{2}h_m\right] \cup \left[-\frac{1}{2}h_m + T, T\right]$; 0, for $x \in [0,T]\backslash I_m$, $M_0^{\nu,m}(x)$ can be defined recurrently by convolution (see (2.1.15)).

Noting that the Fourier expansion of $\overset{\circ}{M}_0^{0,m}(x)$ is

$$\overset{\circ}{M}_0^{0,m}(x) = \sum_{\nu \in \mathbf{Z}} \frac{\sin \nu\pi/K(m)}{\nu\pi} e^{i\frac{2\pi\nu}{T}x}, \qquad (2.1.16)$$

by induction, we have

$$\overset{\circ}{M}_0^{n,m}(x) = \sum_{\nu \in \mathbf{Z}} \left(\frac{\sin \nu\pi/K(m)}{\nu\pi} \right)^{n+1} \exp(i2\pi\nu x/T). \qquad (2.1.17)$$

Now we want to prove

$$\overset{\circ}{B}_0^{n,m}(x) = (K(m))^n \overset{\circ}{M}_0^{n,m}(x) \qquad (2.1.18)$$

Since the restriction of $\overset{\circ}{B}_0^{n,m}(x)$ on $[0,T]$ is $\tilde{B}_0^{n,m}(x)$, and by definition $M_0^{0,m}(x) := \tilde{B}_0^{0,m}(x)$. Therefore, by periodic extension, the formula (2.1.18) is true for $n = 0$.

Suppose $\overset{o}{B}_0^{\nu,m}(x) = (K(m))^\nu\, \overset{o}{M}_0^{\nu,m}(x)$ for some $\nu \geq 0$. Then by (2.1.15)

$$(K(m))^\nu\, \overset{o}{M}_0^{\nu+1,m}(x)$$

$$= \frac{1}{T}\int_0^T \overset{o}{M}_0^{0,m}(x-y)(K(m))^\nu\, \overset{o}{M}_0^{\nu,m}(y)dy$$

$$= \frac{1}{T}\int_0^T \overset{o}{B}_0^{0,m}(x-y)\, \overset{o}{B}_0^{\nu,m}(y)dy$$

$$= \frac{1}{T}\int_{x-\frac{1}{2}h_m}^{x+\frac{1}{2}h_m} \overset{o}{B}_0^{\nu,m}(y)dy$$

$$= \frac{1}{T}\int_{x-\frac{1}{2}h_m}^{x+\frac{1}{2}h_m} \frac{1}{\nu!h_m}\sum_{j=0}^{\nu+1}(-1)^j\binom{\nu+1}{j}$$

$$\times\left[\left(y-\left(j-\frac{\nu+1}{2}\right)h_m\right)_+^\nu\right.$$

$$\left.+\left(y-\left(j+K(m)-\frac{\nu+1}{2}\right)h_m\right)_+^\nu\right]dy$$

$$= \frac{1}{(\nu+1)!h_m^\nu T}\left\{\sum_{j=0}^{\nu+2}(-1)^j\binom{\nu+2}{j}\right.$$

$$\times\left[\left(x-\left(j-\frac{\nu+2}{2}\right)h_m\right)_+^{\nu+1}\right.$$

$$\left.\left.+\left(x-\left(j+K(m)-\frac{\nu+2}{2}\right)h_m\right)_+^{\nu+1}\right]\right\}$$

$$= (K(m))^{-1}\{B_0^{\nu+1,m}(x) + B_{0+K(m)}^{\nu+1,m}(x)\}$$

$$= (K(m))^{-1}\, \overset{o}{B}_0^{\nu+1,m}(x),$$

$$x \in [0,T].$$

Thus $\overset{o}{B}_0^{\nu+1,m}(x) = (K(m))^{\nu+1}\, \overset{o}{M}_0^{\nu+1,m}(x)$, (2.1.18) is true for all $n \in \mathbb{N}$, we have established (2.1.14). $\square$

Theorem 2.1.3   If

$$A_\nu^{n,j}(x) = C_\nu^{n,j}\sum_{l=0}^{K(j)-1}\exp(2\pi i l\nu/K(j))\, \overset{o}{B}_0^{n,j}(x-lh_j) \qquad (2.1.19)$$

then $\{A_\nu^{n,j}\}_{\nu=0}^{K(j)-1}$ is an orthonormal basis for $\tilde{S}_n(h_j)$.

$A_\nu^{n,j}(x)$ has the following Fourier expansion

$$A_\nu^{n,j}(x) = C_\nu^{n,j}(K(j))^{n+1} \sum_{\lambda \in \mathbf{Z}} \left(\frac{\sin \frac{\nu\pi}{K(j)}}{(\nu + \lambda K(j))\pi} \right)^{n+1} \exp\left(\frac{2\pi i(\nu + \lambda K(j))x}{T} \right)$$

$$(2.1.20)$$

where

$$C_\nu^{n,j} = \left[t_0 + 2 \sum_{\lambda=1}^{n} t_\lambda \cos\left(\frac{2\pi\lambda\nu h_j}{T} \right) \right]^{-\frac{1}{2}},$$

$$(2.1.21)$$

$$t_\lambda = B_0^{2n+1}(\lambda, 1), \quad B_0^{2n+1}(\cdot, 1) \in S_{2n+1}(1) \quad \text{(see (2.1.11))}.$$

Proof. The inner product is defined by

$$\langle f, g \rangle := \frac{1}{T} \int_0^t f(y)\overline{g(y)}dy.$$

By (2.1.14), we obtain

$$\langle \overset{\circ}{B_l^{n,j}}(\cdot), \overset{\circ}{B_m^{n,j}}(\cdot) \rangle = (K(j))^{-1} \overset{\circ}{B_0^{2n+1,j}}((m-l)h_j). \qquad (2.1.22)$$

By (2.1.19), (2.1.22), we have

$$\langle A_{k_1}^{n,j}(\cdot), A_{k_2}^{n,j}(\cdot) \rangle = (C_{k_1}^{n,j})^2 \delta_{k_1,k_2} \sum_{\nu=0}^{K(j)-1} \exp\left(\frac{2\pi i\nu k_2}{K(j)} \right) \cdot \overset{\circ}{B_0^{2n+1}}(\nu h_j, h_j).$$

$$(2.1.23)$$

Since

$$\operatorname{supp}\tilde{B}_0^{2n+1}(\cdot, h_j) = [-(n+1)h_j, (n+1)h_j] \cup [T - (n+1)h_j, T + (n+1)h_j],$$

therefore,

$$\tilde{B}_0^{2n+1}(\lambda h_j, h_j) = \begin{cases} B_0^{2n+1}(\lambda h_j, h_j), & \lambda = 0, \cdots, n, \\ 0, & \lambda = n+1, \cdots, K(j) - n - 1, \\ B_0^{2n+1}(\lambda h_j - T, h_j), & \lambda = K(j) - n, \cdots, K(j) - 1. \end{cases}$$

$$(2.1.24)$$

By [P1],

$$B_0^{2n+1}(ah_j, h_j) = B_0^{2n+1}(ah, h)$$

and

$$B_0^{2n+1}(-ah, h) = B_0^{2n+1}(ah, h) = B_0^{2n+1}(a, 1). \qquad (2.1.25)$$

Set

$$
\begin{aligned}
E(n, k, j) : &= \sum_{\lambda=0}^{K(j)-1} \exp\left(\frac{2\pi i k \lambda}{K(j)}\right) \overset{\circ}{B}_0^{2n+1}(\lambda h_j, h_j) \\
&= \sum_{\lambda=0}^{n} \exp\left(\frac{2\pi i k \lambda h_j}{T}\right) B_0^{2n+1}(\lambda, 1) \\
&\quad + \sum_{\lambda=1}^{n} \exp\left(\frac{2\pi i (-\lambda) k h_j}{T}\right) B_0^{2n+1}(\lambda, 1) \quad \text{(by (2.1.24))} \\
&= B_0^{2n+1}(0, 1) + 2 \sum_{\lambda=1}^{n} \cos\left(\frac{2\pi k \lambda h_j}{T}\right) B_0^{2n+1}(\lambda, 1) \\
&= t_0 + 2 \sum_{\lambda=1}^{n} \cos\left(\frac{2\pi k \lambda h_j}{T}\right) t_\lambda = (C_k^{n,j})^{-2},
\end{aligned}
$$

$$(2.1.26)$$

thus $E(n, k, j)(C_k^{n,j})^2 = 1$.

From (2.1.23), we have

$$\langle A_{k_1}^{n,j}(\cdot), A_{k_2}^{n,j}(\cdot) \rangle = \delta_{k_1, k_2}, \quad \text{for} \quad 0 \le k_1, k_2 \le K(j) - 1.$$

Thus $\{A_\nu^{n,j}\}_{\nu=0}^{K(j)-1}$ is an *o.n.* basis of $\tilde{S}_n(h_j)$ since $A_\nu^{n,j} \in \tilde{S}_n(h_j)$ and $\dim \tilde{S}_n(h_j) = K(j)$.

The Fourier expansion of $A_\nu^{n,j}(x)$ follows from (2.1.19) and (2.1.14). $\square$

Since

$$\tilde{S}_n(h_m) \subset \tilde{S}_n(h_{m+1}) \qquad (2.1.27)$$

We would like to ask: what is the expression for $A_\nu^{n,m}(x)$ in the basis of $\tilde{S}_n(h_{m+1})$? The following theorem is the answer to this question.

Theorem 2.1.4 $A_\nu^{n,m}(x)$ satisfies the following two-scale equation:

$$A_\nu^{n,m}(x) = a_\nu^{n,m+1} A_\nu^{n,m+1}(x) + b_\nu^{n,m+1} A_{\nu+K(m)}^{n,m+1}(x) \qquad (2.1.28)$$

where $\{A_\nu^{n,m+1}\}_{\nu=0}^{K(m+1)-1}$ is the *o.n.* basis of $\tilde{S}_n(h_{m+1})$, and

$$a_\nu^{n,m+1} = \frac{C_\nu^{n,m}\left(\cos\dfrac{\nu\pi}{K(m+1)}\right)^{n+1}}{C_\nu^{n,m+1}}, \tag{2.1.29}$$

$$b_\nu^{n,m+1} = \frac{C_\nu^{n,m}\left(\sin\dfrac{\nu\pi}{K(m+1)}\right)^{n+1}}{C_{\nu+K(m)}^{n,m+1}}. \tag{2.1.30}$$

Proof. From the following deduction it is easy to deduce

$$\langle A_\nu^{n,m}, A_\nu^{n,m+1}\rangle = a_\nu^{n,m+1}, \tag{2.1.31}$$

$$\langle A_\nu^{n,m}, A_{\nu+K(m)}^{n,m+1}\rangle = b_\nu^{n,m+1}. \tag{2.1.32}$$

In fact

$$A_\nu^{n,m}(x) = C_\nu^{n,m}(K(m))^{n+1}$$

$$\times \left\{ \sum_{\lambda_1\in\mathbf{Z}} \left[\frac{\sin\dfrac{\nu\pi}{K(m)}}{(\nu+\lambda_1 K(m+1))\pi}\right]^{n+2} \right.$$

$$\times \exp\left(2\pi i\frac{(\lambda_1 K(m+1)+\nu)x}{T}\right)$$

$$+ \sum_{\lambda_2\in\mathbf{Z}} \left[\frac{\sin\dfrac{\nu\pi}{K(m)}}{(\nu+K(m)+\lambda_2 K(m+1))\pi}\right]^{n+1}$$

$$\left. \times \exp\left(2\pi i\frac{(\lambda_2 K(m+1)+\nu+K(m))x}{T}\right) \right\}$$

$$= C_\nu^{n,m}\left(\cos\frac{\nu\pi}{K(m+1)}\right)^{n+1} \frac{A_\nu^{n,m+1}(x)}{C_\nu^{n,m+1}}$$

$$+ C_\nu^{n,m}\left(\sin\frac{\nu\pi}{K(m+1)}\right)^{n+1} \frac{A_{\nu+K(m)}^{n,m+1}(x)}{C_{\nu+K(m)}^{n,m+1}}$$

$$= a_\nu^{n,m+1} A_\nu^{n,m+1}(x) + b_\nu^{n,m+1} A_{\nu+K(m)}^{n,m+1}(x).$$

$\square$

Corollary 2.1.1

$$A_0^{n,m}(x) \equiv 1, \tag{2.1.33}$$

$$C_0^{n,m} = 1, \quad C_{K(j)-\nu}^{n,j} = C_\nu^{n,j}, \quad \nu = 1, \cdots, K(j-1), \tag{2.1.34}$$

$$b_0^{n,m} = 0, \quad a_0^{n,m} = 1. \tag{2.1.35}$$

Proof. From (2.1.26)

$$(C_0^{n,m})^{-2} = E(n,0,m) = \sum_{\lambda=0}^{K(m)-1} \overset{\circ}{B}_0^{2n+1}(\lambda h_j, h_j)$$

$$= \sum_{\lambda=0}^{K(m)-1} \overset{\circ}{B}_0^{2n+1}(-\lambda h_j, h_j) = \sum_{\lambda=0}^{K(m)-1} \tilde{B}_\lambda^{2n+1,j}(0) = 1$$

([P7], (2.1.24) and (2.1.25)). Thus we have (2.1.34). (2.1.33) follows from (2.1.19), (2.1.34) and [P7]. (2.1.35) follows from (2.1.29) and (2.1.30). □.

Definition 2.1.2 The function

$$D_\nu^{n,m}(x) = -b_\nu^{n,m+1} A_\nu^{n,m+1}(x) + a_\nu^{n,m+1} A_{\nu+K(m)}^{n,m+1}(x) \tag{2.1.36}$$

$\nu = 0, \cdots, K(m) - 1$, where $a_\nu^{\nu,m+1}, b_\nu^{n,m+1}$ are defined in (2.1.29) and (2.1.30) respectively, is called the periodic quasi-wavelet. (see p.117 Note 1).

Definition 2.1.3 Given m, $m \in \mathbb{N}$, define two spaces of functions V_m and W_m as follows:

$$V_m := \tilde{S}_n(h_m), \quad W_m := V_{m+1} - V_m, \tag{2.1.37}$$

$W_m \perp V_m$.

Theorem 2.1.5 $\dim W_m = K(m)$, the system of functions $\{D_\nu^{n,m}\}_{\nu=0}^{K(m)-1}$ forms an orthonormal basis of W_m.

Proof. From (2.1.36), $D_\nu^{n,m} \in V_{m+1}$. It is easy to verify the following identities

$$\langle D_{\nu_1}^{n,m}, D_{\nu_2}^{n,m} \rangle = \delta_{\nu_1,\nu_2}, \quad \langle D_{\nu_1}^{n,m}, A_{\nu_2}^{n,m} \rangle = 0 \qquad (2.1.38)$$

for any $0 \le \nu_1, \nu_2 \le K(m) - 1$.

From the second identity of (2.1.38), $D_\nu^{n,m} \notin V_m$, hence $D_\nu^{n,m} \in W_m$, for $\nu = 0, \cdots, K(m) - 1$ and from first identity of (2.1.38), $\{D_\nu^{n,m}\}_0^{K(m)-1}$ is an o.n. basis of W_m, since $\dim W_m = \dim V_{m+1} - \dim V_m = K(m+1) - K(m) = K(m)$. □.

Let $\overset{\circ}{L}_2(0,T)$ be the space of periodic square integrable functions.

Let P_m, Q_m be the projection operators which maps $\overset{\circ}{L}_2(0,T)$ onto V_m and W_m respectively. For $f \in \overset{\circ}{L}_2(0,T)$, define

$$\alpha_\nu^m = \langle f, A_\nu^{n,m} \rangle, \quad \beta_\nu^m = \langle f, D_\nu^{n,m} \rangle. \qquad (2.1.39)$$

Theorem 2.1.6   The coefficients α_ν^m and β_ν^m in (2.1.39) satisfy the decomposition formula

$$\begin{pmatrix} \alpha^m \\ \beta^m \end{pmatrix} = \begin{pmatrix} L_{m+1} \\ H_{m+1} \end{pmatrix} \begin{pmatrix} \alpha^{m+1} \\ \beta^{m+1} \end{pmatrix} \quad \text{(refer to (2.1.48))}, \qquad (2.1.40)$$

and the reconstruction formula

$$\begin{pmatrix} \alpha^{m+1} \\ \beta^{m+1} \end{pmatrix} = \begin{pmatrix} L_{m+1}^T, H_{m+1}^T \end{pmatrix} \begin{pmatrix} \alpha^m \\ \beta^m \end{pmatrix} \quad \text{(refer to (2.1.45))}, \qquad (2.1.41)$$

where $\alpha^m = (\alpha_0^m, \cdots, \alpha_{K(m)-1}^m)^T, \beta^m = (\beta_0^m, \cdots, \beta_{K(m)-1}^m)^T$, $(\cdot)^T$ denotes the transpose of $(\cdot)$, L_m and H_m are the upper and lower half of the matrix M_m respectively, and

$$M_m = \begin{pmatrix} a_0^{n,m} & 0 & \cdots & b_0^{n,m} & 0 & \cdots \\ 0 & a_1^{n,m} & 0 & 0 & b_1^{n,m} & 0 & \cdots \\ & & \cdots & & & \cdots \\ & & \cdots & a_{K(m)-1}^{n,m} & & \cdots & b_{K(m)-1}^{n,m} \\ b_0^{n,m} & 0 & \cdots & -a_0^{n,m} & 0 & \cdots \\ 0 & b_1^{n,m} & 0 & 0 & -a_1^{n,m} & 0 & \cdots \\ & & \cdots & & & \cdots \\ & & \cdots & b_{K(m)-1}^{n,m} & \cdots & 0 & \cdots & -a_{K(m)-1}^{n,m} \end{pmatrix} \qquad (2.1.42)$$

M_m is of $K(m) \times K(m)$ matrix.

Proof.   Let $f \in \overset{\circ}{L}_2 (0,T)$. Since $P_{m+1}f \in V_{m+1}$, then

$$(P_{m+1}f)(x) = \sum_{\nu=0}^{K(m+1)-1} \alpha_\nu^{m+1} A_\nu^{n,m+1}(x). \tag{2.1.43}$$

On the other hand, $V_{m+1} = V_m \oplus W_m$. Hence

$$(P_m f)(x) + (Q_m f)(x)$$

$$= \sum_{\nu=0}^{K(m)-1} [\alpha_\nu^m A_\nu^{n,m}(x) + \beta_\nu^m D_\nu^{n,m}(x)]$$

$$= \sum_{\nu=0}^{K(m)-1} \left\{ \alpha_\nu^m [a_\nu^{n,m+1} A_\nu^{n,m+1}(x) + b_\nu^{n,m+1} A_{\nu+K(m)}^{n,m+1}(x)] \right.$$

$$\left. + \beta_\nu^m [-b_\nu^{n,m+1} A_\nu^{n,m+1}(x) + a_\nu^{n,m+1} A_{\nu+K(m)}^{n,m+1}(x)] \right\}$$

(from (2.1.28) and (2.1.36))

$$= \sum_{\nu=0}^{K(m)-1} \left\{ (a_\nu^{n,m+1} \alpha_\nu^m - b_\nu^{n,m+1} \beta_\nu) A_\nu^{n,m+1}(x) \right.$$

$$\left. + (b_\nu^{n,m+1} \alpha_\nu^m + a_\nu^{n,m+1} \beta_\nu^m) A_{\nu+K(m)}^{n,m+1}(x) \right\}. \tag{2.1.44}$$

Comparing the coefficients of both sides of (2.1.43) and (2.1.44), we have

$$\alpha_\nu^{m+1} = a_\nu^{n,m+1} \alpha_\nu^m - b_\nu^{n,m+1} \beta_\nu, \quad \nu = 0, \cdots, K(m) - 1$$

$$\alpha_{\nu+K(m)}^{m+1} = b_\nu^{n,m+1} \alpha_\nu^m + a_\nu^{n,m+1} \beta_\nu^m, \quad \nu = 0, \cdots, K(m) - 1. \tag{2.1.45}$$

(2.1.45) is the reconstruction formula (2.1.41).

From (2.1.28) and the orthonormality of the system $\{A_\nu^{n,m}\}_{\nu=0}^{K(m)-1}$, we can easily prove

$$|a_\nu^{n,m}|^2 + |b_\nu^{n,m}|^2 = 1, \quad \nu = 0, \cdots, K(m) - 1. \tag{2.1.46}$$

Hence

$$M_m M_m^T = I_m, \quad M_m^{-1} = M_m^T \tag{2.1.47}$$

where I_m is the unit matrix of size $K(m) \times K(m)$. From (2.1.41) and (2.1.47) we obtain the following decomposition formulas:

$$\alpha_\nu^m = a_\nu^{n,m+1} \alpha_\nu^{m+1} + b_\nu^{n,m+1} \alpha_{\alpha+K(m)}^{m+1}, \tag{2.1.48}$$
$$\beta_\nu^m = b_\nu^{n,m+1} \alpha_\nu^{m+1} - a_\nu^{n,m+1} \alpha_{\nu+K(m)}^{m+1}, \quad \nu = 0, \cdots, K(m) - 1.$$

if we write (2.1.48) in the form of matrix, it is just (2.1.40).   □.

In the following we shall give the error estimates.

Define the space of continuous periodic functions on $I = [0, T]$ by

$$C^0(I) = \{f \in C(I) : f(T) = f(0)\}. \tag{2.1.49}$$

Similarly, we define the space of m-times continuously differentiable periodic functions by

$$\overset{\circ}{C}{}^m(I) = \{f \in C^m(I) : D^j f(T) = D^j f(0), j = 0, \cdots, m\}, \tag{2.1.50}$$

$$L_p^\sigma(I) = \{f : D^{\sigma-1}f \in AC(I) \quad \text{and} \quad D^\sigma f \in L_p(I)\}. \tag{2.1.51}$$

We define this periodic Sobolev space by

$$\overset{\circ}{L}{}_p^\sigma(I) = L_p^\sigma(I) \cap \overset{\circ}{C}{}^{\sigma-1}(I) \tag{2.1.52}$$

and introduce the periodic modulus of smoothness

$$\overset{\circ}{\omega}_r(f,t)_p = \sup_{0 < h \le t} \| \Delta_h^r f \|_{L_p(I)}.$$

where $\Delta_h^r f(t) = r! h^r [t, t+h, \cdots, t+rh]f$ is the r-th forward difference of f at t.

Since

$$\|f - P_m f\|_2 = \inf_{s \in V_m} \|f - s\|_2$$

(cf. [Dav, p.172]), thus we have the following theorem (cf. [Sc, p.307–308]):

Theorem 2.1.7   (i) There exists a constant C_1 such that for all $f \in \overset{\circ}{L}_2$ $(0,T)$,

$$\|f - P_m f\|_2 \le C_1 \overset{\circ}{\omega}_{n+1}(f, h_m)_2. \tag{2.1.53}$$

(ii) There exists constant C_2 such that for all $f \in \overset{\circ}{L}_2^\sigma (0,T)$ $(1 \leq \sigma \leq n+1)$,

$$\|f - P_m f\|_2 \leq C_2 (h_m)^\sigma \overset{\circ}{\omega}_{n+1-\sigma} (D^\sigma f, h_m)_2. \tag{2.1.54}$$

From (2.1.53), it is clear that

$$\lim_{m \to \infty} \|f - P_m f\|_2 = 0 \tag{2.1.55}$$

for any $f \in \overset{\circ}{L}_2 (0,T)$.

We conclude that $\{V_m\}_{m \geq 0}$ is dense in $\overset{\circ}{L}_2 (0,T)$. From (2.1.54), the smoother f is, the faster $\|f - P_m f\|_2$ tends to zero when m approaches infinity.

Remark   The above methods of construction the periodic scaling functions and quasi-wavelets have been extended to two dimensional case [CLJ].

§2.2   Quasi-Wavelets on the Unit Circle

At the end of Chapter I we have pointed out in Note 1 that complex splines on the circle play an essential role in complex approximation and in the approximation of conformal mapping. But we have only introduced the interpolation and quasi-interpolant there. Here we intend to develop the complex approximation by combining the idea of multiresolution analysis.

In Chapter I (§1.1) we have defined the class of functions $\Phi(\triangle)$, where $\triangle = \{z_1, \cdots, z_K | z_\nu \in \Gamma\}$. Now let z_ν be the point ω_l^ν, $\omega_l^\nu := \exp(i\nu h_l)$, $h_l = 2\pi / K(l)$, $K(l) = 2^l K, K \geq n+2, l \in \mathbb{N}$. $\triangle_l := \{\omega_l^\nu | \nu = 0, \cdots, K(l) - 1\}$. Thus $\Phi_n(\triangle_l)$ is the class of complex splines on Γ of degree n with knots $\triangle_l$.

Define

$$V_l := \Phi_n(\triangle_l). \tag{2.2.1}$$

Then $\{N_{j,n}^{(l)}\}_{j=0}^{K(l)-1}$ is a basis of V_l (see (1.1.3), §1.1, [P2]). $N_{j,n}^{(l)}(e^{i\theta})$ has the Fourier expansion.

$$N_{j,n}^{(l)}(l^{i\theta}) = K_n^{(l)} \sum_{\nu \in \mathbb{Z}} C_\nu^{(l)} \cdots C_{\nu-n}^{(l)} e^{i\nu(\theta - jh_l)}, \tag{2.2.2}$$

(see (1.1.9)).

Theorem 2.2.1　If $P_{k,n}^{(l)}(z)$ $(k = 0, \cdots, K(l) - 1)$ are given by.

$$P_{k,n}^{(l)}(z) = A_{k,n}^{(l)} \sum_{\nu=0}^{K(l)-1} \omega_l^{\nu k} N_{\nu,n}^{(l)}(z), \quad z \in \Gamma \tag{2.2.3}$$

then $\{P_{k,n}^{(l)}\}_{k=0}^{K(l)-1}$ is a orthonormal basis of V_l, where

$$A_{k,n}^{(l)} = \left[K(l) \sum_{\nu=0}^{K(l)-1} \langle N_{0,n}^{(l)}, N_{\nu,n}^{(l)} \rangle \omega_l^{-k\nu} \right]^{-\frac{1}{2}}.$$

Proof.　Without loss of generality, we assume $l = 0$. Simplifying the notation, for instance by setting $P_{k,n} = P_{k,n}^{(0)}$, we obtain by (2.2.2) that

$$\langle P_{k_1,n}, P_{k_2,n} \rangle : = \frac{1}{2\pi} \int_0^{2\pi} P_{k_1,n}(e^{ix}) \overline{P_{k_2,n}(e^{ix})} dx$$

$$= M \cdot A_{k_1,n} \overline{A}_{k_2,n} \sum_{\nu_1=0}^{K-1} \sum_{\nu_2=0}^{K-1} \omega^{\nu_1 k_1 - \nu_2 k_2}$$

where

$$M = \langle N_{\nu_1,n}(\cdot), N_{\nu_2,n}(\cdot) \rangle = |K_n|^2 \sum_{\lambda \in \mathbf{Z}} |C_\lambda C_{\lambda-1} \cdots C_{\lambda-n}|^2 \omega^{\lambda(\nu_2 - \nu_1)}.$$

Substituting into the above expression, we have

$$\langle P_{k_1,n}, P_{k_2,n} \rangle = |A_{k_1,n}|^2 |K_n|^2 \delta_{k_1,k_2} \sum_{\eta \in \mathbf{Z}} |C_{k_1+\eta K} \cdots C_{k_1+\eta K-n}|^2 = \delta_{k_1,k_2}.$$

It is clear that $A_{k,n}^{(l)} > 0$, since

$$\sum_{\nu=0}^{K-1} \langle N_{0,n}, N_{\nu,n} \rangle \omega^{-k\nu} = |K_n|^2 K \sum_{\eta \in \mathbf{Z}} |C_{k+\eta K} \cdots C_{k+\eta K-n}|^2$$

$$> 0.$$

□.

Remark   Before establishing Theorem 2.2.1, we did not know what is the *o.n.* basis of the space of complex spline functions. Now we can apply multi-resolution analysis to the space $\Phi_n(\triangle_l)$.

We define the function $M_{\nu,n}^{(l)}$ by formula (1.1.7), and now establish the two scale equation in the following

Lemma 2.2.1   For any $l \in \mathbb{N}, M_{\nu,n}^{(l)}(z)$ satisfies the following two scale equation:

$$M_{\nu,n}^{(l)}(z) = \sum_{j=0}^{n+1} a_{j,n}^{(l+1)} M_{j+2\nu,n}^{(l+1)}(z), \quad \nu = 0, \cdots, K(l) - 1, \quad z \in \Gamma \quad (2.2.4)$$

where the coefficients $a_{j,n}^{(l+1)}$ can be calculated from the Taylor expansion of the function $\prod_{k=0}^{n}(\omega_{l+1}^k z + 1) = \sum_{j=0}^{n+1} a_{j,n}^{(l+1)} z^j$, or, can be calculated from the recursion formula:

$$a_{0,n}^{(l+1)} = a_{0,0}^{(l+1)} = a_{1,0}^{(l+1)} = 1, \quad a_{n+1,n}^{(l+1)} = \omega_{l+1}^n a_{n,n-1}^{(l+1)}$$

$$a_{j,n}^{(l+1)} = \omega_{l+1}^j a_{j,n-1}^{(l+1)} + \omega_{l+1}^{j-1} a_{j-1,n-1}^{(l+1)}, \quad 1 \le j \le n. \quad (2.2.5)$$

Proof.   From §1.1,

$$\mathrm{supp} M_{\nu,0}^{(l)} = \gamma(\omega_l^\nu, \omega_l^{\nu+1}),$$

and

$$M_{\nu,0}^{(l)}(z) = M_{2\nu,0}^{(l+1)}(z) + M_{2\nu+1,0}^{(l+1)}(z).$$

Thus (2.2.4) is proved for $n = 0$, and we have

$$a_{0,0}^{(l+1)} = a_{1,0}^{(l+1)} = 1.$$

The rest can be proved by (1.1.6), (1.1.7) and induction.   □.

The Fourier expansion of $P_{k,n}^{(l)}(z)$ is

$$P_{k,n}^{(l)}(z) = \sum_{\lambda \in \mathbb{Z}} g_{k,n}^{(l)}(\lambda) e^{i(k+\lambda K(l))x}, \quad z = e^{ix},$$

$$g_{k,n}^{(l)}(\lambda) = K(l) K_n^{(l)} A_{k,n}^{(l)} \prod_{\nu=0}^{n} C_{k+\lambda K(l)-\nu}^{(l)}, \quad (2.2.6)$$

where $K_n^{(l)}, A_{k,n}^{(l)}$ are constants in (2.2.2) and (2.2.3) respectively.

The following theorem gives the two scale equation for $P_{k,n}^{(l)}(z)$:

Theorem 2.2.2 There exist constants $\hat{a}_{j,n}^{(l+1)}, j = k, k + K(l)$, such that

$$P_{k,n}^{(l)}(z) = \hat{a}_{k,n}^{(l+1)} P_{k,n}^{(l+1)}(z) + \hat{a}_{k+K(l),n}^{(l+1)} P_{k+K(l),n}^{(l+1)}(z), \quad z \in \Gamma \qquad (2.2.7)$$

where

$$\hat{a}_{k,n}^{(l+1)} = \langle P_{k,n}^{(l)}, P_{k,n}^{(l+1)} \rangle$$

and

$$\hat{a}_{k+K(l),n}^{(l+1)} = \langle P_{k,n}^{(l)}, P_{k+K(l)}^{(l+1)} \rangle, \quad k = 0, \cdots, K(l) - 1; l \in \mathbb{N}.$$

Proof. Without loss of generality, we may suppose $l = 0$. From (2.2.3), (1.1.8), (2.2.4) and (1.1.9) we obtain

$$P_{k,n}(z) = \frac{K_n A_{k,n}}{(2\pi i)^n} \sum_{\nu=0}^{K-1} \omega^{\nu k} \sum_{j=0}^{n+1} a_{j,n}^{(1)} M_{j,n}^{(1)}(ze^{-i\nu h}) \qquad (2.2.8)$$

Since $\{M_{k,n}^{(1)}\}_0^{K(1)-1}$ and $\{P_{k,n}^{(1)}\}_0^{K(1)-1}$ are bases of V_1, there are constants $\{d_\lambda\}_0^{K(1)-1}$ such that

$$M_{j,n}^{(1)}(z) = \sum_{\lambda=0}^{K(1)-1} d_\lambda^j P_{\lambda,n}^{(1)}(z), \quad j = 0, \cdots, K(1) - 1. \qquad (2.2.9)$$

Substituting (2.2.9) into (2.2.8), we obtain

$$P_{k,n}(z) = \frac{K_n A_{k,n} K}{(2\pi i)^n} \sum_{j=0}^{n+1} a_{j,n}^{(1)}$$

$$\times \left\{ d_k^j \sum_{\eta \in \mathbb{Z}} g_{k,n}^{(1)}(\eta) e^{i(k+\eta K(1))x} + d_{k+K}^j \sum_{\eta \in \mathbb{Z}} g_{k+K,n}^{(1)}(\eta) e^{i(k+K+\eta K(1))x} \right\}$$

$$= \hat{a}_{k,n}^{(1)} P_{k,n}^{(1)}(z) + \hat{a}_{k+K,n}^{(1)} P_{k+K,n}^{(1)}(z),$$

here we used the formula (2.2.6). The rest of Theorem follow from the orthonormality of $\{P_{k,n}^{(1)}\}_0^{K(1)-1}$.   □

Multiplying $\overline{P}_{k,n}^{(l)}(z)$ to both sides of (2.2.7) and then integrating we obtain

$$|\hat{a}_{k,n}^{(l+1)}|^2 + |\hat{a}_{k+K(l),n}^{(l+1)}|^2 = 1 \qquad (2.2.10)$$

Now we construct the quasi-wavelets on Γ.

From Theorem 2.2.2, it follows that

$$V_0 \subset V_1 \subset \cdots V_l \subset \cdots, \quad \dim V_l = K(l). \qquad (2.2.11)$$

Define $Q_{k,n}^{(l)}(z)$ as follows:

$$Q_{k,n}^{(l)}(z) = -\overline{\hat{a}}_{k+K(l),n}^{(l+1)} P_{k,n}^{(l+1)}(z) + \overline{\hat{a}}_{k,n}^{(l+1)} P_{k+K(l),n}^{(l+1)}(z). \qquad (2.2.12)$$

It is easy to check $\langle Q_{k_1,n}^{(l)}, Q_{k_2,n}^{(l)} \rangle = \delta_{k_1,k_2}$ for $0 \le k_1, k_2 \le K(l) - 1$ and $\langle P_{k,n}^{(l)}, Q_{j,n}^{(l)} \rangle = 0$ for $0 \le k, j \le K(l) - 1$. Denote the orthogonal complement of V_l in V_{l+1} by W_l. We have

Theorem 2.2.3   The set of functions $\{Q_{k,n}^{(l)}\}_0^{K(l)-1}$ is an orthonormal basis of W_l. Moreover, $V_{l+1} = V_l \oplus W_l$.

From (2.2.7) and (2.2.12) we have

$$
\begin{aligned}
P_{k,n}^{(l+1)} &= \overline{\hat{a}}_{k,n}^{(l+1)} P_{k,n}^{(l)} - \hat{a}_{k+K(l),n}^{(l+1)} Q_{k,n}^{(l)}, \quad k = 0, \cdots, K(l) - 1 \\
P_{k+K(l),n}^{(l+1)} &= \overline{\hat{a}}_{k+K(l),n}^{(l+1)} P_{k,n}^{(l)} + \hat{a}_{k,n}^{(l+1)} Q_{k,n}^{(l)}, \quad k = 0, \cdots, K(l) - 1.
\end{aligned}
\qquad (2.2.13)
$$

The functions $\{Q_{k,n}^{(l)}\}_{k=0}^{K(l)-1}$, for any $l \in \mathbb{N}$, are called the circular quasi-wavelets on the unit circle.

Denote the space of all square integrable functions on Γ by $\overset{\circ}{L}_2(\Gamma)$. P_l, Q_j are orthogonal projectors which maps $\overset{\circ}{L}_2(\Gamma)$ onto V_l and W_j respectively.

Theorem 2.2.4   For $c_k^{(l)} = \langle P_l f, P_{k,n}^{(l)} \rangle, d_k^{(l)} = \langle Q_l f, Q_{k,n}^{(l)} \rangle$, we have the decomposition formulas

$$
\begin{aligned}
c_k^{(l)} &= \overline{\hat{a}}_{k,n}^{(l+1)} c_k^{(l+1)} + \overline{\hat{a}}_{k+K(l),n}^{(l+1)} c_{k+K(l)}^{(l+1)}, \quad k = 0, \cdots, K(l) - 1 \\
d_k^{(l)} &= -\hat{a}_{k+K(l),n}^{(l+1)} c_k^{(l+1)} + \hat{a}_{k,n}^{(l+1)} c_{k+K(l)}^{(l+1)}, \quad k = 0, \cdots, K(l) - 1
\end{aligned}
\qquad (2.2.14)
$$

and the reconstruction formulas

$$c_k^{(l+1)} = \hat{a}_{k,n}^{(l+1)} c_k^{(l)} - \overline{\hat{a}}_{k+K(l),n}^{(l+1)} d_k^{(l)}, \quad k = 0, \cdots, K(l) - 1$$

$$c_{k+K(l)}^{(l+1)} = \hat{a}_{k+K(l),n}^{(l+1)} c_k^{(l)} + \overline{\hat{a}}_{k,n}^{(l+1)} d_k^{(l)}, \quad k = 0, \cdots, K(l) - 1$$

(2.2.15)

where

$$\hat{a}_{k,n}^{(l+1)} = \frac{A_{k,n}^{(l)}}{A_{k,n}^{(l+1)}} \cdot \omega_{l+1}^{-\frac{(n+1)k}{2}} \cos(kh_{l+2}) \prod_{\nu=1}^{n} \left(\frac{\cos(\nu - k)h_{l+2}}{\cos \nu h_{l+2}} \right) \quad (2.2.16)$$

$$\hat{a}_{k+K(l),n}^{(l+1)} = \frac{A_{k,n}^{(l)}}{A_{k+K(l),n}^{(l+1)}} \cdot \omega_{l+1}^{-\frac{(n+1)(k+K(l))}{2}}$$

$$\times \cos((k + K(l)h_{l+2}) \prod_{\nu=1}^{n} \left(\frac{\cos(\nu - k - K(l))h_{l+2}}{\cos \nu h_{l+2}} \right)$$

(2.2.17)

$k = 0, \cdots, K(l) - 1$, where

$$A_{j,n}^{(m)} = \left\{ K(m) \sum_{\nu=0}^{K(m)-1} e^{-ij\nu h_m} \langle N_{0,n}^{(m)}, N_{\nu,n}^{(m)} \rangle \right\}^{-\frac{1}{2}}.$$

Proof. Since

$$c_k^{(l)} = \langle P_l f, P_{k,n}^{(l)} \rangle = \langle P_l f + Q_l f, P_{k,n}^{(l)} \rangle$$

$$= \langle P_{l+1} f, P_{k,n}^{(l)} \rangle$$

$$= \sum_{\lambda=0}^{K(l+1)-1} c_\lambda^{(l+1)} \langle P_{\lambda,n}^{(l+1)}, P_{k,n}^{(l)} \rangle,$$

using (2.2.13) we obtain the first formula in (2.2.14). Similarly, from definition

$$d_k^{(l)} = \langle Q_l f, Q_{k,n}^{(l)} \rangle$$

$$= \langle P_{l+1} f, Q_{k,n}^{(l)} \rangle$$

$$= \sum_{\lambda=0}^{K(l+1)-1} c_\lambda^{(l+1)} \langle P_{\lambda,n}^{(l+1)}, Q_{k,n}^{(l)} \rangle,$$

by (2.2.13) the second formula in (2.2.14) is immediate.

Since

$$c_k^{(l+1)} = \langle P_{l+1} f, P_{k,n}^{(l+1)} \rangle = \langle P_l f + Q_l f, P_{k,n}^{(l+1)} \rangle,$$

by (2.2.13) we have the first expression (2.2.15). The second formula in (2.2.15) can be derived similarly.

From (2.2.7), (2.2.3), (1.1.8) and (2.2.4), we have

$$\xi_{\nu_1,\nu_2}^{(l)} := \langle N_{\nu_1,n}^{(l)}(\cdot), N_{\nu_2,n}^{(l+1)}(\cdot) \rangle$$

$$= \frac{K_n^{(l)}}{K_n^{(l+1)}} \sum_{j=0}^{n+1} a_{j,n}^{(l+1)} |K_n^{(l+1)}|^2 \qquad (2.2.18)$$

$$\times \sum_{\lambda \in \mathbb{Z}} \prod_{\eta=0}^{n} |C_{\lambda-\eta}^{(l+1)}|^2 \omega_{l+1}^{\lambda(\nu_2-j-2\nu_1)},$$

$$\hat{a}_{k,n}^{(l+1)} = \langle P_{k,n}^{(l)}, P_{k,n}^{(l+1)} \rangle$$

$$= A_{k,n}^{(l)} A_{k,n}^{(l+1)} \sum_{\nu_1=0}^{K(l)-1} \sum_{\nu_2=0}^{K(l+1)-1} \omega_l^{\nu_1 k} \omega_{l+1}^{-\nu_2 k} \xi_{\nu_1,\xi_2}^{(l)}$$

$$= A_{k,n}^{(l)} A_{k,n}^{(l+1)} \frac{K_n^{(l)}|K_n^{(l+1)}|^2}{K_n^{(l+1)}} \sum_{j=0}^{n+1} a_{j,n}^{(l+1)} \sum_{\lambda \in \mathbb{Z}} \prod_{\eta=0}^{n} |C_{\lambda-\eta}^{(l+1)}|^2 \omega_{l+1}^{-\lambda j}$$

$$\times \sum_{\nu_1=0}^{K(l)-1} \omega_l^{\nu_1(k-\lambda)} \cdot \sum_{\nu_2=0}^{K(l+1)-1} \omega_{l+1}^{\nu_2(\lambda-k)} \quad \text{(Lemma 2.2.1)}$$

$$= A_{k,n}^{(l)} A_{k,n}^{(l+1)} \frac{K_n^{(l)}|K_n^{(l+1)}|^2}{K_n^{(l+1)}} \sum_{\lambda \in \mathbb{Z}} \left(\prod_{\eta=0}^{n} |C_{\lambda-\eta}^{(l+1)}|^2 \right) \prod_{\nu=0}^{n} (\omega_{l+1}^{\nu-\lambda} + 1)$$

$$\times \sum_{\nu_1=0}^{K(l)-1} \omega_l^{\nu_1(k-\lambda)} \cdot \sum_{\nu_2=0}^{K(l+1)-1} \omega_{l+1}^{\nu_2(\lambda-k)} \quad \text{((2.2.18) and (2.2.2))}$$

$$= A_{k,n}^{(l)} A_{k,n}^{(l+1)} \frac{K_n^{(l)}|K_n^{(l+1)}|^2}{K_n^{(l+1)}} \sum_{\lambda \in \mathbb{Z}} \left(\prod_{\eta=0}^{n} |C_{k+\lambda K(l+1)-\eta}^{(l+1)}|^2 \right)$$

$$\times \left(\prod_{\nu=0}^{n} (\omega_{l+1}^{\nu-k} + 1) \right) K(l) K(l+1)$$

$$= A_{k,n}^{(l)} A_{k,n}^{(l+1)} \frac{K_n^{(l)}|K_n^{(l+1)}|^2}{K_n^{(l+1)}} (A_{k,n}^{(l+1)})^{-2} (K(l+1)|K_n^{(l+1)}|)^{-2}$$

$$\times K(l) K(l+1) \prod_{\nu=0}^{n} (\omega_{l+1}^{\nu-k} + 1)$$

$$= \frac{A_{k,n}^{(l)}}{A_{k,n}^{(l+1)}} \cdot \frac{K_n^{(l)}}{K_n^{(l+1)}} \cdot \frac{K(l)}{K(l+1)} \cdot \left[\prod_{\nu=0}^{n} \cos\left(\frac{\nu-k}{2} h_{l+1} \right) \right] 2^{n+1} \omega_{l+1}^{\frac{(n+1)}{2}(\frac{n}{2}-k)}$$

$$= \frac{A_{k,n}^{(l)}}{A_{k,n}^{(l+1)}} \cdot \frac{K(l)}{2K(l)} \cdot \omega_{l+1}^{-\frac{n}{4}(n+1)} 2^{-n} \left[\prod_{\nu=1}^{n} \cos\left(\frac{\nu}{2}h_{l+1}\right) \right]^{-1}$$

$$\times 2^{n+1} \omega_{l+1}^{\frac{n}{4}(n+1) - \frac{k(n+1)}{2}} \left[\prod_{\nu=0}^{n} \cos\left(\frac{\nu - k}{2}h_{l+1}\right) \right]$$

$$= \frac{A_{k,n}^{(l)}}{A_{k,n}^{(l+1)}} \omega_{l+1}^{-\frac{(n+1)k}{2}} \cos(kh_{l+2}) \prod_{\nu=1}^{n} \left[\frac{\cos(\nu - k)h_{l+2}}{\cos(\nu h_{l+2})} \right], \quad \text{this is (2.2.16)}.$$

By an analogous approach, we have (2.2.17).   □.

In §1.2 we gave the error estimates for a function in $\overset{\circ}{C}{}^n$ by the quasi-interpolant, but there is a function space $\overset{\circ}{L}{}_p^{n+1}(I)$ such that

$$\overset{\circ}{C}{}^{n+1}(I) \subseteq \overset{\circ}{L}{}_p^{n+1}(I) \subseteq \overset{\circ}{C}(I), \quad \text{for} \quad 1 \le p \le \infty. \tag{2.2.19}$$

(see (2.1.51)).

In the following, we give an error estimate for a function in $\overset{\circ}{L}{}_2^{n+1}(I)$ by the quasi-interpolant spline in V_0. It will be useful in the proof of Theorem 2.2.6.

Theorem 2.2.5   Let $f \in \overset{\circ}{L}{}_2^{n+1}(I)$, and $\mathcal{L}(f)$ be the quasi-interpolant spline function of degree n as defined in (1.2.1). Then there is a constant C_0 independent of f and h, such that

$$\|f - \mathcal{L}(f)\|_2 \le C_0 h^{n+\frac{1}{2}} \|f^{(n+1)}\|_2$$

where $\|f\|_2 = \|f\|_{\overset{\circ}{L}{}_2(I)}, h = 2\pi/K$ and $f^{(n+1)} = \dfrac{d^{n+1}f}{dz^{n+1}}.$ (2.2.20)

Proof.   Let

$$Y_z f \ = \sum_{r=0}^{n} \frac{f^{(r)}(z)(\cdot - z)^r}{r!} \in \pi_n,$$

$$f \ = Y_z f + R_z,$$

$$E(z) \ = f - \mathcal{L}(f).$$

We have

$$E^{(s)}(z) = f^{(s)}(z) - (\mathcal{L}(Y_z f - R_z))^{(s)}(z) = -(\mathcal{L}(R_z))^{(s)}(z) \quad 0 \le s \le n, z = e^{ix}. \tag{2.2.21}$$

From $R_z(t) = f(t) - Y_z f(t)$, we have

$$\frac{d^n}{dt^n} R_z(t) = f^{(n)}(t) - f^{(n)}(z).$$

On the other hand, the r-th derivative of the remainder of the Taylor expansion $R_z(t)$ is

$$R_z^{(r)}(t) = \frac{1}{(n-1-r)!} \int_z^t (R_z(y))^{(n)} (t-y)^{n-1-r} dy$$

$$= \frac{1}{(n-1-r)!} \int_z^t [f^{(n)}(y) - f^{(n)}(z)](t-y)^{n-1-r} dy \qquad (2.2.22)$$

$$(0 \leq r \leq n-1)$$

where the integral path is along Γ. Then, from (2.2.21) and (1.2.1)

$$|E(z)| \leq |\mathcal{L}(R_z)(z)| \leq \sum_{j=0}^{K-1} |L_j(R_z)||N_{j,n}(z)|$$

$$\leq \sum_{j=0}^{K-1} \sum_{r \leq n} |T_{j,r}||R_z^{(r)}(t_j)||N_{j,n}(z)|$$

$$\leq \sum_{j=0}^{K-1} \left\{ \sum_{r \leq n-1} \frac{|T_{j,r}|}{(n-r-1)!} \left| \int_z^{t_j} \left(\int_z^y f^{(n+1)}(t)dt \right) (t_j - y)^{n-1-r} dy \right| \right.$$

$$\left. + |T_{j,n}| \left| \int_z^{t_j} f^{(n+1)}(t)dt \right| \right\} |N_{j,n}(z)|.$$

There exists $\eta_j \in \gamma(z, t_j)$, such that the right hand side of above inequality is less than the following quantity:

$$\sum_{j=0}^{K-1} \left\{ \frac{\pi}{2} \sum_{r \leq n-1} \frac{T_{j,r}}{(n-r-1)!} \left| \int_z^{\eta_j} f^{(n+1)}(t)dt \right| \cdot |t_j - z|^{n-r} \right.$$

$$\left. + |T_{j,n}| \left| \int_z^{t_j} f^{(n+1)}(t)dt \right| \right\} |N_{j,n}(z)|$$

$$= \sum_{j=0}^{K-1} \left\{ \sum_{r \leq n-1} \left| \int_z^{\eta_j} f^{(n+1)}(t)dt \right| \Gamma_{n-r,j} + \left| \int_z^{t_j} f^{(n+1)}(t)dt \right| \Gamma_{0,j} \right\}$$

$$(2.2.23)$$

where

$$\left.\begin{array}{l}
\Gamma_{n-r,j}(z) = \frac{\pi}{2} \frac{|T_{j,r}|}{(n-1-1)!} |t_j - z|^{n-r} |N_{j,n}(z)|, \quad 0 \le r \le n-1 \\[2mm]
\Gamma_{0,j}(z) = |T_{j,n}||N_{j,n}(z)|
\end{array}\right\}.$$

$$(2.2.24)$$

From Lemma 1.2.1, (b) and the definition of $T_{j,r}$ (§1.2), we have $0 \le \Gamma_{r,j} \le Ch^n, r = 0, \cdots, n$, where C is a constant.

Substitute (2.2.24) into (2.2.23)

$$\|E\|_2 = \left\{ \frac{1}{2\pi} \int_0^{2\pi} |E(e^{ix})|^2 dx \right\}^{\frac{1}{2}} \quad (z = e^{ix})$$

$$\le Ch^n \left\{ \frac{1}{2\pi} \sum_{\lambda=0}^{K-1} \int_{\lambda h}^{(\lambda+1)h} \left[\sum_{j=\lambda-n}^{\lambda} \sum_{r \le n-1} \left| \int_z^{\eta_j} f^{(n+1)}(t) dt \right| \right. \right. \quad (2.2.24)_1$$

$$\left. \left. + \left| \int_z^{t_j} f^{(n+1)}(t) dt \right| \right]^2 dx \right\}^{\frac{1}{2}}$$

here we used the fact: supp $N_{j,n}(z) = \gamma(\omega^j, \omega^{j+n+1}) = \text{supp}\Gamma_{r,j}(z)$, and

$$|R_z^{(r)}(t_j)| \le \max_{\eta \in \bar{\gamma}_{z,t_j}} |f^{(n)}(\eta) - f^{(n)}(z)||t_j - z|^{n-1-r}|\gamma_{z,t_j}| \frac{1}{(n-r-1)!}$$

$$= |f^{(n)}(\eta_j) - f^{(n)}(z)||t_j - z|^{n-1-r}|\gamma_{z,t_j}| \frac{1}{(n-r-1)!}$$

$$\le |f^{(n)}(\eta_j) - f^{(n)}(z)||t_j - z|^{n-r} \cdot \frac{\pi}{2} \cdot \frac{1}{(n-r-1)!}$$

where $|\gamma_{a,b}|$ denotes the length of arc $\gamma_{a,b}$ on Γ. For any fixed $\lambda, 0 \le \lambda \le K-1$, there are $n+1$ $N_{j,n}$, such that $(\text{supp}N_{j,n}) \cap \gamma(\omega^\lambda, \omega^{\lambda+1}) \ne \phi$, for $j = \lambda - n, \cdots, \lambda$. By Hölder's inequality the right side of $(2.2.24)_1$ is less than or equal to $C_0 h^{n+\frac{1}{2}} \|f^{(n+1)}\|_2$. $\qquad \square$.

Lemma 2.2.2 Let

$$\mathcal{L}^M := D^m + \sum_{j=0}^{m-1} a_j(x) D^j$$

be a linear differential operator where $a_j \in \overset{\circ}{C}^{m-j}([a,b]), j = 0, \cdots, m-1$. and let its null space be $\ker(\mathcal{L}^m) := \{f | f \in \overset{\circ}{L}_1^m([a,b]), \mathcal{L}^m f(x) = 0$ for all $x \in [a, b]$}. Then there exists a constant C_1, depending on ker $(\mathcal{L}^m)$, such that for all intervals $J \subseteq [a, b]$ of length $|J| < (b - a)/2$, and for all $u \in$ ker $(\mathcal{L}^m)$, we have

$$\|D^j u\|_{\overset{\circ}{L_p}(J)} \leq C_1 |J|^{-j + \frac{1}{p} - \frac{1}{q}} \|u\|_{\overset{\circ}{L_q}(J)} \tag{2.2.25}$$

where $1 \leq p, q \leq \infty$.

The proof of this lemma is analogue to that of Theorem 10.2 in [Sc, p.421] where the theorem was proved for the non-periodic case.

Let f be any function in $\overset{\circ}{L}_2 (I)$. Define

$$P_l(f)(z) = \sum_{k=0}^{K(l)-1} \langle f, P_{k,n}^{(l)} \rangle P_{k,n}^{(l)}(z)$$

as the orthogonal projection of f into V_l, where P_l is the orthogonal projector. Given $\varepsilon > 0$, if we request: $\|f - P_l(f)\|_2 < \varepsilon$, how large the number l should be ? The following theorem will give an answer to this problem.

Theorem 2.2.6   Let P_l be the orthogonal projector from $\overset{\circ}{L}_2 (I)$ onto $V_l, I = [0, 2\pi]$. Given any f in $\overset{\circ}{L}_2 (I)$ and any $\varepsilon > 0$, there is an integer l_0, such that

$$\|f - P_l(f)\|_2 < \varepsilon, \quad \text{for} \quad l \geq l_0 \tag{2.2.26}$$

where

$$l_0 = \ln\{h[C_2(1 + 2\|f\|_2/\varepsilon)]^{\frac{1}{n + \frac{1}{2}}}\} \frac{2}{\ln},$$

and C_2 as in (2.2.30).

Proof.   Set

$$S_N(x) = \sum_{\nu = -N}^{N} \hat{f}(\nu) e^{i\nu x},$$

where

$$\hat{f}(\nu) = \frac{1}{2\pi} \int_0^{2\pi} f(x) e^{-i\nu x} dx.$$

Given $\varepsilon > 0$, there exists N_0, for all $N \geq N_0$, such that (see [R, p.92])

$$\|f - S_N\|_2 < \frac{\varepsilon}{2}. \tag{2.2.27}$$

From Theorem 2.2.5, there is a function $\mathcal{L}(S_{N_0})$ in V_l, such that

$$\|S_{N_0} - \mathcal{L}(S_{N_0})\|_2 \le C_0 h_l^{n+\frac{1}{2}} \|S_{N_0}^{(n+1)}\|_2 \qquad (2.2.28)$$

where C_0 is independent of h_l and S_{N_0}. In the light of the following formula

$$S_{N_0}^{(n+1)}(x) = \frac{d^{n+1}S_{N_0}(x)}{dz^{n+1}} = z^{-n-1}\sum_{j=1}^{n+1} a_j D_x^j S_{N_0}(x)$$

we have

$$\|S_{N_0}^{(n+1)}\|_2 \le \sum_{j=1}^{n+1} |a_j| \|D_x^j S_{N_0}\|_2. \qquad (2.2.29)$$

Since $S_{N_0} \in \ker(\mathcal{L}^m)$, where $\mathcal{L}^m := (D^2 + N_0^2)\cdots(D^2+1)D, m = 2N_0+1$, applying Lemma 2.2.2 to the case $[a,b] = [0, 2\pi], J = \left[0, \frac{\pi}{4}\right], p = q = 2$, we obtain the following estimate

$$\|S_{N_0}^{(n+1)}\|_2 \le C_1 \sum_{j=1}^{n+1} |a_j| \left(\frac{4}{\pi}\right)^j \left(\frac{\varepsilon}{2} + \|f\|_2\right).$$

By (2.2.28) and above inequality

$$\|S_{N_0} - \mathcal{L}(S_{N_0})\|_2 \le C_2 h_l^{n+\frac{1}{2}} \left(\frac{\varepsilon}{2} + \|f\|_2\right), \qquad (2.2.30)$$

where

$$C_2 = C_0 C_1 \max_{1 \le j \le n+1} |a_j| \sum_{k=1}^{n+1} \left(\frac{4}{\pi}\right)^k.$$

Therefore, for all $l \ge l_0 + 1$

$$\begin{aligned}
\inf_{S \in V_l} \|f - S\|_2 = \|f - P_l(f)\|_2 \\
\le \|f - \mathcal{L}(S_{N_0})\|_2 \\
\le \|f - S_{N_0}\|_2 + \|S_{N_0} - \mathcal{L}(S_{N_0})\|_2 \\
\le \frac{\varepsilon}{2} + C_2 h_l^{n+\frac{1}{2}} \left(\frac{\varepsilon}{2} + \|f\|_2\right) \\
\le \varepsilon.
\end{aligned}$$

$\square$

As for the smooth functions, we have the following error estimate:

Theorem 2.2.7   Let $f^{(n+1)}$ be continuous on Γ. For any $\varepsilon > 0$, and $l \geq r_0$, we have

$$\|f - P_l f\|_2 < \varepsilon \tag{2.2.31}$$

where

$$r_0 = \left\{ \frac{1}{n+1} \ln\left(\frac{P_0 \|f^{(n+1)}\|_\infty}{\varepsilon} \right) + \ln h \right\} \frac{2}{\ln}, \tag{2.2.32}$$

$$P_0 < \frac{\pi}{4}(n+1)(n+2)[2n(n+1)]^n \frac{1}{n!}. \tag{2.2.33}$$

Proof.

$$\|f - P_l f\|_2 \leq \|f - \mathcal{L}(f)\|_2 \leq P_0 h_l^{n+1} \|f^{(n+1)}\|_\infty$$

the last inequality follows from Theorem 1.2.4 and the fact:

$$\|E\|_2 = \left\{ \frac{1}{2\pi} \int_0^{2\pi} |E(e^{ix})|^2 dx \right\}^{\frac{1}{2}} \tag{2.2.34}$$

$$\leq P_0 \|f^{(n+1)}\|_\infty h_l^{n+1}.$$

The inequality (2.2.33) follows from (1.2.14) and (1.2.17) for $s = 0$.

Now if $l \geq r_0$, then from (2.2.34), we have (2.2.31).   □

(cf. *p.117* Note 2.)

§2.3   Antiperiodic Orthonormal Quasi-Wavelets

In previous sections we have introduced the periodic orthonormal bases for some function space. In other problems we also encounter functions with different boundary conditions.

In this section, we consider the anti-periodic function space.

A function $f(x)$ is called anti-periodic π if $f(x + \pi) = -f(x), \forall x \in R^1$. We define the space of spline functions

$$V_m^a := \{f \mid f^{(j)}(x + \pi) = -f^{(j)}(x), j = 0, \cdots, n-1;$$

$$f(x) \in \pi_n \text{ if } x \in [y_\nu^m, y_{\nu+1}^m), f \in C^{n-1}(J)\}, \tag{2.3.1}$$

where $J = [0, \pi]$, $y_j^m = y_0^m + jh_m, j \in \mathbb{Z}$, $y_0^m = -n_0 h_m, h_m = h/3^m$, $h = \pi/L$, $n_0 = \frac{n+1}{2}$, $L \in \mathbb{Z}_+, n \in \mathbb{N}$.

We also use the B-spline functions as in formula (2.1.11). The only change is that here we employ the length of step h_m, $h_m := h/3^m$, instead of $h/2^m$, and change T to 2π. We can also define the periodic (with period 2π) B-spline function $\tilde{B}_i^{n,m}$ and its periodic extension $\overset{\circ}{B}_i^{n,m}$.

$$\tilde{B}_i^{n,m}(x) = B_i^{n,m}(x) + B_{i+K(m)}^{n,m}(x), \quad x \in [0, 2\pi], \quad K(m) = 3^m K. \quad (2.3.2)$$

$i = 0, \cdots, K(m) - 1$. $\{\tilde{B}_i^{n,m}\}_0^{K(m)-1}$ forms a basis for $\tilde{S}_n(h_m)$ (see (§2.1)).

Analogous to (2.1.19) and (2.1.20), we define

$$A_{k,3}^{n,m}(x) = C_k^{n,m} K^{n+1}(m) \sum_{\nu \in \mathbb{Z}} \left(\frac{\sin(\nu + \frac{k}{K(m)})\pi}{(\nu K(m) + k)\pi} \right)^{n+1} \exp(i(k + \nu K(m))x),$$

$$(2.3.3)$$

$k = 0, \cdots, K(m) - 1$, where

$$K(m) = 3^m K, \quad C_k^{n,m} = \left[t_0 + 2 \sum_{\lambda=1}^n t_\lambda \cos(\lambda \nu h_j) \right]^{-\frac{1}{2}},$$

$$h_j = h/3^j, \quad t_\lambda = B_0^{2n+1}(\lambda, 1).$$

If we define $V_m := \tilde{S}_n(h_m) = \text{span}\{\tilde{B}_i^{n,m}(x)|i = 0, \cdots, K(m) - 1, x \in [0, 2\pi]\}$, then $V_m \subset V_{m+1}$. But the two scale equations are different from (2.1.28), in fact, we have the following

Theorem 2.3.1 $\{A_{\nu,3}^{n,m}\}_{\nu=0}^{K(m)-1}, \{A_{\nu,3}^{n,m+1}\}_{\nu=0}^{K(m+1)-1}$ are the o.n. bases in the spaces V_m and V_{m+1} respectively. We have the following 3-scale equations

$$A_{\nu,3}^{n,m}(x) = \sum_{\lambda=0}^2 a_{\nu,\nu+\lambda K(m)} A_{\nu+\lambda K(m),3}^{n,m+1}(x), \quad \nu = 0, \cdots, K(m) - 1. \quad (2.3.4)$$

where

$$a_{\nu,\nu+\lambda K(m)} = \langle A_{\nu,3}^{n,m}, A_{\nu+\lambda K(m),3}^{n,m+1} \rangle, \quad \lambda = 0, 1, 2. \quad (2.3.5)$$

Proof.   It is easy to check the orthonormality. We can prove (2.3.4) by two different approaches:

(a) Using (2.3.3) directly, since for any $\nu \in \mathbf{Z}$ and $k \in \{0, 1, \cdots, K(m) - 1\}$, we have the following decomposition formulas:

$$\nu = \lambda + 3\mu, \quad \lambda \in \{0, 1, 2\}, \quad \mu \in \mathbf{Z}.$$

Then

$$k + \nu K(m) = \begin{cases} k + \mu K(m+1), & \lambda = 0 \\ (k + K(m)) + \mu K(m+1), & \lambda = 1 \\ (k + 2K(m)) + \mu K(m+1), & \lambda = 2. \end{cases} \tag{2.3.6}$$

Substituting (2.3.6) into (2.3.3), regrouping again, we obtain (2.3.4).

(b) By using the following formula:

$$\tilde{B}_l^{n,m}(x) = \sum_{k=-n-1}^{n+1} p_{n,k} \tilde{B}_{k+3l}^{n,m+1}(x) \tag{2.3.7}$$

where $\tilde{B}_l^{n,m}(x)$ is defined in (2.3.2) and

$$p_{n,k} = \begin{cases} 3^{-n} \displaystyle\sum_{\nu=0}^{[\frac{n+1-k}{2}]} \binom{n+1}{\nu}\binom{n+1-k-\nu}{\nu}, & k = -(n+1), \cdots, n+1 \\ 0, & \text{otherwise} \end{cases}$$

(2.3.7) is also true for the extension $\overset{\circ}{B}_l^{n,m}(x)$, then we use (2.1.19) to obtain (2.3.4).

$$\square.$$

We note that in formula (2.3.4), the right side involves three terms.

Definition 2.3.1   Let

$$a_{i,j} = \langle A_{i,3}^{n,m}, A_{j,3}^{n,m+1} \rangle,$$

$$q_k = \left[1/(1 - |a_{k,k+K(m)}|^2) \right]^{\frac{1}{2}}.$$

Define

$$D_{j,3}^{n,m}(x) =$$
$$\begin{cases} -\bar{a}_{j,j+K(m)} \cdot a_{j,j} \cdot q_j A_{j,3}^{n,m+1}(x) + \frac{1}{q_j} A_{j+K(m),3}^{n,m+1}(x) \\[2mm] \quad -\bar{a}_{j,j+K(m)} \cdot a_{j,j+2K(m)} q_j A_{j+2K(m),3}^{n,m+1}(x), \quad 0 \le j \le K(m)-1 \\[2mm] -\bar{a}_{j-K(m),j+K(m)} q_{j-K(m)} A_{j-K(m),3}^{n,m+1}(x) \\[2mm] +\bar{a}_{j-K(m),j-K(m)} \cdot q_{j-K(m)} A_{k+K(m),3}^{n,m+1}(x) \quad K(m) \le j \le 2K(m)-1. \end{cases}$$
$$(2.3.8)$$

and define

$$W_m := \text{the orthogonal complementary space of } V_m \text{ in } V_{m+1}.$$

Theorem 2.3.2 $\{D_{j,3}^{n,m}(x)\}_{j=0}^{2K(m)-1}$ is an o.n. basis for W_m, and $V_{m+1} = V_m \oplus W_m$, i.e.

$$\langle D_{j_1,3}^{n,m}, D_{j_2,3}^{n,m} \rangle = \delta_{j_1,j_2}, \quad 0 \le j_1, j_2 \le 2K(m)-1$$

and

$$\langle D_{j,3}^{n,m}, A_{k,3}^{n,m} \rangle = 0, \quad 0 \le j \le 2K(m)-1, 0 \le k \le K(m)-1$$

Proof. Since

$$\langle A_{j_1,3}^{n,m+1}, A_{j_2,3}^{n,m+1} \rangle = \delta_{j_1,j_2}$$

and

$$\langle A_{j,3}^{n,m}, A_{j,3}^{n,m} \rangle = |a_{j,j}|^2 + |a_{j,j+K(m)}|^2 + |a_{j,j+2K(m)}|^2 = 1,$$

the following inequalities are easily verified

$$\langle A_{j,3}^{n,m}, D_{l,3}^{n,m} \rangle = q_j a_{j,j+K(m)} \left[\frac{1}{q_j^2} - |a_{j,j}|^2 - |a_{j,j+2K(m)}|^2 \right] \delta_{j,l} = 0 \cdot \delta_{j,l} = 0$$

for $0 \le j, l \le K(m)-1$,

$$\langle A_{j,3}^{n,m}, D_{l+K(m),3}^{n,m} \rangle = [-a_{j,j} \cdot a_{j,j+2K(m)} q_j + a_{j,j} q_j a_{j,j+2K(m)}] \delta_{j,l}$$
$$= 0 \cdot \delta_{j,l} = 0$$

for $0 \leq j, l \leq K(m) - 1$,

$$\langle D_{j,3}^{n,m}, D_{l+K(m),3}^{n,m} \rangle$$

$$= (a_{j,j}q_j^2\bar{a}_{j,j+K(m)}a_{j,j+2K(m)} - a_{j,j}q_j^2\bar{a}_{j,j+K(m)}a_{j,j+2K(m)}) \cdot \delta_{j,l} = 0$$

for $0 \leq j, l \leq K(m) - 1$,

$$\langle D_{j,3}^{n,m}, D_{k,3}^{n,m} \rangle = (|a_{j,j+K(m)}|^2 + 1 - |a_{j,j+K(m)}|^2)\delta_{j,k}$$
$$= \delta_{j,k}.$$

for $0 \leq j, k \leq K(m) - 1$,

$$\langle D_{j+K(m),3}^{n,m}, D_{k+K(m),3}^{n,m} \rangle = (1 - |a_{j,j+K(m)}|^2)q_j^2\delta_{j,k} = \delta_{j,k}$$

for $0 \leq j, k \leq K(m) - 1$. $\qquad\qquad\qquad\qquad\qquad\qquad$ □.

Since $\{A_{0,3}^{n,m+1}, A_{1,3}^{n,m+1}, \cdots, A_{K(m+1)-1,3}^{n,m+1}\}$ and $\{A_{0,3}^{n,m}, \cdots, A_{K(m)-1,3}^{n,m},$ $D_{0,3}^{n,m}, \cdots, D_{2K(m)-1,3}^{n,m}\}$ are orthonormal bases in V_{m+1}, there is a unitary matrix which transforms the former to the latter. This transform is presented in the formulas (2.3.4) and (2.3.8).

Based on the results given above, we can define the scaling function in V_m^a and the quasi-wavelet in the antiperiod case.

At the beginning of this section we have introduced a positive integer L, such that $\pi = hL$, thus $K = 2L$. Now let $L \geq n+1$, $L(m) = 3^m L$.

Definition 2.3.2 $\overset{\circ}{E}_i^{n,m}(x) := \overset{\circ}{B}_i^{n,m}(x) - \overset{\circ}{B}_{i+L(m)}^{n,m}(x)$, where $\overset{\circ}{B}_i^{n,m}$ (x) is the 2π periodic extension of $\tilde{B}_i^{n,m}(x)$ (see (2.3.2)).

In the following, for simplicity, let n be odd: $n = 2n_0 + 1$, $n_0 \in \mathbb{N}$. We now prove a basic theorem for the space V_m^a (refer to the beginning of §2.3).

Lemma 2.3.1 $\{\overset{\circ}{E}_i^{n,m}\}_{i=0}^{L(m)-1}$ is a basis of the space V_m^a, where $E_i^{n,m}(x)$ is the restriction of $\overset{\circ}{E}_i^{n,m}(x)$ on $J := [0, \pi]$.

Proof. For $0 \leq j \leq n-1$, by 2π periodicity of the function $\overset{\circ}{B}_i^{n,m}(x)$, we have

$$D^j \overset{o}{E_i^{n,m}}(x+\pi) = D^j\{\overset{o}{B_i^{n,m}}(x+\pi) - \overset{o}{B_{i+L(m)}^{n,m}}(x+\pi)\}$$

$$= D^j\{\overset{o}{B_i^{n,m}}(x-\pi) - \overset{o}{B_{i+L(m)}^{n,m}}(x+\pi)\}$$

$$= D^j\{\overset{o}{B_i^{n,m}}(x-L(m)h_m) - \overset{o}{B_{i+L(m)}^{n,m}}(x+L(m)h_m)\}$$

$$= D^j\{\overset{o}{B_{i+L(m)}^{n,m}}(x) - \overset{o}{B_i^{n,m}}(x)\}$$

$$= -D^j E_i^{n,m}(x),$$

therefore $\overset{o}{E_i^{n,m}}(x)$ is of antiperiod π, and $E_i^{n,m}(x) \in V_m^a$, for $i = 0, \cdots,$ $L(m) - 1$.

Assume there are constants $\{C_i\}_{-n_0}^{-n_0-1+L(m)}$, such that

$$C_{-n_0}E_{-n_0}^{n,m}(x) + \cdots + C_{-n_0-1+L(m)}E_{-n_0-1+L(m)}^{n,m}(x) = 0, \quad x \in J. \quad (2.3.9)$$

Set

$$a_i = \begin{cases} C_i, & i = -n_0, \cdots, -n_0 - 1 + L(m) \\ -C_{i-L(m)}, & i = -n_0 + L(m), \cdots, -n_0 - 1 + K(m). \end{cases}$$

Then, from (2.3.9) and the antiperiodicity of $E_j^{n,m}(x)$ (for $j = -n_0, \cdots,$ $-n_0 - 1 + L(m)$), we have

$$\sum_{i=-n_0}^{-n_0-1+K(m)} a_i \overset{o}{B_i^{n,m}}(x) \equiv 0, \quad x \in [0, 2\pi].$$

Therefore $a_i = 0$ for all i since $\{\overset{o}{B_i^{n,m}}\}_{i=-n_0}^{-n_0-1+K(m)}$ is a linearly independent system. This leads to the independency of the system $\{E_j^{n,m}, j = -n_0, \cdots, -n_0 - 1 + L(m)\}$. Since the dimension of V_m^a is $L(m)$, Lemma 2.3.1 is proved.   □.

Definition 2.3.3   The inner product on $J = [0, \pi]$ is defined by

$$(f, g) = \frac{1}{\pi}\int_0^\pi f(x)\overline{g(x)}dx.$$

Lemma 2.3.2   If f and g are of antiperiod π, then

$$\frac{1}{\pi}\int_0^\pi f(x)\overline{g(x)}dx = \frac{1}{2\pi}\int_0^{2\pi} f(x)\overline{g(x)}dx \qquad (2.3.10)$$

namely,

$$(f,g) = \langle f,g\rangle.$$

Proof.

$$\langle f,g\rangle = \frac{1}{2\pi}\left\{\int_0^\pi + \int_\pi^{2\pi} f(x)\overline{g(x)}dx\right\}$$

$$= \frac{2}{2\pi}\int_0^\pi f(x)\overline{g(x)}dx$$

$$= (f,g).$$

$\square$.

Definition 2.3.4 $A_{n,m}^{a,j}(x) := \frac{1}{2}[A_{2j-1,3}^{n,m}(x) - A_{2j-1,3}^{n,m}(x-\pi)]$ where $A_{\nu,3}^{n,m}(x)$ is defined in (2.3.3).

Theorem 2.3.3 $\{A_{n,m}^{a,j}\}_{j=0}^{L(m)-1}$ is an o.n. basis in V_m^a.

Proof.   From (2.3.3), we have

$$A_{\nu,3}^{n,m}(x - kh_m) = \exp(-i\nu kh_m)A_{\nu,3}^{n,m}(x) \qquad (2.3.11)$$

and further, by periodicity,

$$\langle A_{2j_1-1,3}^{n,m}(\cdot), A_{2j_2-1,3}^{n,m}(\cdot+\pi)\rangle = \langle A_{2j_1-1,3}^{n,m}(\cdot-\pi), A_{2j_2-1,3}^{n,m}(\cdot)\rangle$$

$$= -\delta_{j_1,j_2}.$$

Since $A_{n,m}^{a,j}$ is antiperiodic with antiperiod π, for $0 \le j_1, j_2 \le L(m) - 1$, and we have

$$(A_{n,m}^{a,j_1}, A_{n,m}^{a,j_2}) = \langle A_{n,m}^{a,j_1}, A_{n,m}^{a,j_2}\rangle$$

$$= \frac{1}{4}\{2\delta_{j_1,j_2} - 2(-1)^{2j_1-1}\delta_{j_1,j_2}\}$$

$$= \delta_{j_1,j_2}.$$

$A_{n,m}^{a,j} \in V_m^a$, $\{A_{n,m}^{a,j}\}_{j=0}^{L(m)-1}$ is an orthonormal basis of V_m^a.   □.

The complementary subspace of V_m^a in V_{m+1}^a is denoted by W_m^a, and

$$\dim W_m^a = \dim V_{m+1}^a - \dim V_m^a$$

$$= (3^{m+1} - 3^m)L = 2L(m).$$

Definition 2.3.5 $D_{n,m}^{a,j}(x) := \frac{1}{2}(D_{2j-1,3}^{n,m}(x) - D_{2j-1,3}^{n,m}(x - \pi))$ where $D_{l,3}^{n,m}(x)$ is defined in (2.3.8).

Theorem 2.3.4 The system of functions $\{D_{n,m}^{a,j}(x)\}_{j=1}^{2L(m)}$ constitutes an o.n. basis of W_m^a, i.e.

$$(D_{n,m}^{a,j_1}, D_{n,m}^{a,j_2}) = \delta_{j_1,j_2}, \quad 1 \leq j_1, j_2 \leq 2L(m) \qquad (2.3.12)$$

and

$$(D_{n,m}^{a,j}, A_{n,m}^{a,l}) = 0, \quad 1 \leq j \leq 2L(m), \quad 0 \leq l \leq L(m) - 1 \qquad (2.3.13)$$

thus

$$V_{m+1}^a = W_m^a \oplus V_m^a, \quad W_m^a \perp V_m^a. \qquad (2.3.14)$$

Proof. From the 2π periodicity of the function $D_{2j-1,3}^{n,m}(x)$, we can derive $D_{n,m}^{a,j}(x + \pi) = -D_{n,m}^{a,j}(x)$. By Definition 2.3.5, $D_{n,m}^{a,j}(x) \in V_{m+1}^a$ for $j = 1, \cdots, 2L(m)$.

$(D_{n,m}^{a,j_1}(\cdot), D_{n,m}^{a,j_2}(\cdot))$

$= \frac{1}{4}\{\langle D_{2j_1-1,3}^{n,m}(\cdot), D_{2j_2-1,3}^{n,m}(\cdot)\rangle + \langle D_{2j_1-1,3}^{n,m}(\cdot - \pi), D_{2j_2-1,3}^{n,m}(\cdot - \pi)\rangle$

$- \langle D_{2j_1-1,3}^{n,m}(\cdot), D_{2j_2-1,3}^{n,m}(\cdot - \pi)\rangle - \langle D_{2j_1-1,3}^{n,m}(\cdot - \pi), D_{2j_2-1,3}^{n,m}(\cdot)\rangle$

$= \frac{1}{4}\{2\delta_{j_1,j_2} - 2(-1)^{2j_1-1}\delta_{j_1,j_2}\} = \delta_{j_1,j_2}.$

Formula (2.3.13) can be easily proved by Definition 2.3.5 and Definition 2.3.4.

□.

Using the functions $A_{n,m}^{a,j}$ and $D_{n,m}^{a,j}$, we can establish the reconstruction and decomposition formulas. Each of them involves only 3 terms. (see p.117 Note 3).

§2.4 Real Valued Periodic Quasi-wavelets

In this section we consider only real valued functions, thus a large class of functions are covered. For instance, the classes of various real-valued spline functions are involved. (cf. p.117 Note 4).

We will also show that in some cases, the scaling functions C_α^j, S_α^j converge to cosine and sine functions respectively.

Different from the quasi-wavelets considered in previous sections, here, each of the decomposition and reconstruction formulas involves 4 terms.

Let a compactly supported real-valued function $\varphi(x) \in L_2(R^1)$ satisfy the following conditions:

(i)   for some $p \in Z^+, 2p < N := K/2$, where K is a positive even integer, and $h = T/K$. The support of φ : supp$\varphi \subset [-ph, ph]$,

(ii) φ is refinable, i.e. there exists $\{C_k\} \in l^2$, such that

$$\varphi(x) = \sum_{k \in Z} c_k \varphi(2x - kh), \qquad (2.4.1)$$

(iii) $\int_R \varphi(x)dx \neq 0,$ \hfill (2.4.2)

(iv)   The functions $\varphi(x - lh)$ for $l = -p+1, \cdots, K+p-1$ are linearly independent on $[0, T]$.

Denote the 2^j dilation of φ by φ^j where

$$\varphi^j(x) = \varphi(2^j x). \qquad (2.4.3)$$

The T-periodization of φ is denoted by Φ_α^j

$$\Phi_\alpha^j(x) = \sum_{\lambda \in Z} \varphi^j(x + \lambda T - \alpha h_j), \quad x \in R \qquad (2.4.4)$$

where $\alpha \in \mathbb{Z}, j \in \mathbb{N}$.

From (2.4.1) and (2.4.4), we can derive

$$\Phi^j_\lambda(x) = \sum_{k=-p}^{p} c_k \Phi^{j+1}_{k+2\lambda}(x) \quad \text{for} \quad x \in [0, T] \tag{2.4.5}$$

where $\{c_k\}$ are the coefficients in (2.4.1).

Define

$$V_j = \text{span}\{\Phi^j_\alpha : \alpha = 0, 1, \cdots, K(j) - 1\}. \tag{2.4.6}$$

From (2.4.5), $V_j \subseteq V_{j+1}$. Moreover, we have the following:

Theorem 2.4.1   Under the assumptions (i)–(iv), we have:

(a) $\dim V_j = K(j)$,

(b) $\overline{U_{j \geq 0} V_j} = \overset{\circ}{L}_2(0, T)$.

Proof.   Suppose there exist constants $\{C_l\}$, such that

$$\sum_{l=0}^{K(j)-1} C_l \Phi^j_l(x) = 0 \quad x \in [0, T]. \tag{2.4.7}$$

By (2.4.4) and (2.4.3), and setting $2^j x = yh$, we have

$$\sum_{l=0}^{K(j)-1} C_l \sum_{\lambda \in \mathbb{Z}} \varphi(yh + \lambda 2^j Kh - lh) = 0, \quad y \in [0, K(j)]. \tag{2.4.8}$$

From assumption (refer to (i)), for fixed $\lambda, \varphi(yh + \lambda K(j)h - lh)$ should vanish if the following inequalities do not satisfy

$$-p < y + \lambda K(j) - l < p. \tag{2.4.9}$$

Now we consider several cases:

(1) $0 < y < 1$. From (2.4.9) we have

$$y + \lambda K(j) - p < l < y + \lambda K(j) + p. \tag{2.4.10}$$

If $\lambda \leq -1$, then $l \leq p - K(j) < 0$, but this is impossible since $l \geq 0$. Thus $\lambda \geq 0$. If $\lambda \geq 2$, then $l \geq 1 - p + 2K(j) > K(j)$, this is impossible since $l \leq K(j) - 1$. Thus the only possible cases are $\lambda = 0$ and $\lambda = 1$.

For $\lambda = 0$, (2.4.10) is $y - p < l < y + p, l = 0, 1, \cdots, p$.

For $\lambda = 1$, from the left side of (2.4.10), $l > y - p + K(j)$, it leads to $l \geq 1 - p + K(j)$, thus $l = K(j) - p + 1, \cdots, K(j) - 1$.

Therefore, from (2.4.8)

$$\sum_{l=0}^{p} C_l \varphi(yh - lh) + \sum_{l=1-p}^{-1} C_{l+K(j)} \phi(yh - lh) = 0, \quad 0 < y < 1.$$

From assumption (iv) we have

$$C_l = 0 \quad \text{for} \quad l = 0, 1, \cdots, p \quad \text{and} \quad l = K(j) - p + 1, \cdots, K(j) - 1. \quad (2.4.11)$$

Substitute (2.4.11) into (2.4.8). (2.4.8) becomes

$$\sum_{l=p+1}^{K(j)-p} C_l \sum_{\lambda \in \mathbf{Z}} \varphi(yh + \lambda 2^j Kh - lh) = 0, \quad y \in [0, K(j)]. \quad (2.4.12)$$

(2) $\nu < y < \nu + 1, 1 \leq \nu \leq K(j) - 1$. Then $y = \nu + \varepsilon, 0 < \varepsilon < 1$. From (2.4.9)

$$\nu + \varepsilon - p + \lambda K(j) < l < \nu + \varepsilon + p + \lambda K(j). \quad (2.4.13)$$

If $\lambda \geq 2$, then there is a contradiction: $K(j) - 1 \geq l > 2K(j) + \nu + \varepsilon - p > K(j)$. If $\lambda \leq -2$, then $l < \nu + p + \varepsilon - 2K(j) < 0$. But this is impossible, since $l \geq 0$. If $\lambda = -1$, then from (2.4.13), $l \leq \nu + p - K(j) \leq K(j) - 1 + p - K(j) = p - 1$, but from (2.4.12), $l \geq p + 1$. Thus the only cases are $\lambda = 0$ and $\lambda = 1$. For $\lambda = 0$, since $\nu + \varepsilon - p < l < \nu + p + \varepsilon, l = \nu - p + 1, \cdots, \nu + p$. For the case $\lambda = 1$, from (2.4.13), $l \geq \nu + 1 - p + K(j) \geq K(j) + 2 - p$ but from (2.4.12), $l \leq K(j) - p$, a contradiction. This asserts that when $\lambda = 1$, no such l exists. Thus the only possible case is $\lambda = 0$, and the equation (2.4.12) is now

$$\sum_{l=p+1}^{\nu+p} C_l \phi(yh - lh) = 0, \quad yh \in (\nu h, (\nu + 1)h), \quad 1 \leq \nu \leq K(j) - 1. \quad (2.4.14)$$

From assumption (iv) we have

$$C_l = 0, \quad l = p+1, \cdots, \nu+p, \quad \text{if} \quad 1 \le \nu \le K-1. \tag{2.4.15}$$

Now equation (2.4.12) can be reduced into

$$\sum_{l=p+K}^{K(j)-p} C_l \sum_{\lambda \in \mathbf{Z}} \varphi(yh + \lambda 2^j K h - lh) = 0, \quad y \in [0, K(j)]. \tag{2.4.16}$$

From (2.4.15), the remaining cases that need to be considered are those $\nu : K \le \nu \le K(j) - 1$. Note that in this case equation (2.4.14) can be written into

$$\sum_{l=K+p}^{\nu+p} C_l \varphi(yh - lh) = 0, \quad y \in (\nu, \nu+1), \quad K \le \nu \le K(j) - 1. \tag{2.4.17}$$

It is easy to prove $C_{K+p} = 0$ by using assumption (iv) and $\nu = K$, in (2.4.17). By induction we can prove $C_l = 0$ for $l \ge K + p + 1$. Thus, from (2.4.11), (2.4.15) and the results just obtained above, we conclude that $C_l = 0$ for $l = 0, \cdots, K(j) - 1$. Thus, from (2.4.7), $\{\Phi_l^j(\cdot)\}_{l=0}^{K(j)-1}$ is an independent class of functions in V_j, from (2.4.6), we have $\dim V_j = K(j)$.

Now we start to prove (b).

By (2.4.4), evidently,

$$\Phi_\alpha^j(x - h_j) = \Phi_{\alpha+1}^j(x), \quad \Phi_{K(j)-1}^j(x - h_j) = \Phi_0^j(x). \tag{2.4.18}$$

Denote the union of all V_j for $j \ge 0$ by $V = \bigcup_{j \ge 0} V_j$. For any j, by (2.4.18), V_j is a h_j-shift invariant space. If $f \in V$, then there is $j_0 \in \mathbf{N}$, such that $f \in V_{j_0}$, and $f(x - h_{j_0}) \in V_{j_0}$.

Suppose that $g(x) \in V^\perp$. Then for any $f \in V$, there is j, such that

$$\langle f(\cdot - \lambda h_j), g \rangle = \langle f, g \rangle := \frac{1}{T} \int_0^T f(x)\overline{g(x)}dx = 0,$$

for any $\lambda \in \mathbf{Z}$. Let g and f have the following Fourier expansion

$$f(x) = \sum_\mu S_\mu \exp(-2\pi i\mu x/T),$$

and

$$g(x) = \sum_\lambda \eta_\lambda \exp(-2\pi i \lambda x / T)$$

respectively.

It follows

$$0 = \langle f(x - lh_j), g(x) \rangle$$

$$= \sum_{\nu=0}^{K(j)-1} \sum_{\mu \in \mathbb{Z}} S_{\nu+\mu K(j)} \overline{\eta}_{\nu+\mu K(j)} \exp\left(\frac{2\pi i \nu l h_j}{T}\right).$$

Multiplying both sides by $\exp(-2\pi i \tilde{\nu} l h_j / T)$ and summing up over index l, we obtain

$$\sum_{\mu \in \mathbb{Z}} S_{\tilde{\nu}+\mu K(j)} \overline{\eta}_{\tilde{\nu}+\mu K(j)} = 0, \quad \text{for} \quad j > 0, \tilde{\nu} = 0, \cdots, K(j) - 1. \quad (2.4.19)$$

For any fixed $\nu, \nu \in \mathbb{Z}$, we therefore have

$$S_\nu \overline{\eta}_\nu = 0. \quad (2.4.20)$$

(2.4.20) is obtained from the absolute convergence of the series $\sum_{\mu \in \mathbb{Z}} S_\mu \overline{\eta}_\mu$ and (2.4.19). In fact, from (2.4.19),

$$S_\nu \overline{\eta}_\nu = - \sum_{\mu \in \mathbb{Z}, \mu \neq 0} S_{\nu+\mu K(j)} \overline{\eta}_{\mu+\nu K(j)}, \quad \nu \in \mathbb{Z}$$

tends to zero as j tends to infinity.

Putting $f(x) = \Phi_0^j(x)$, then for any $\nu \in \mathbb{Z}$

$$0 = S_\nu \overline{\eta}_\nu = \overline{\eta}_\nu \cdot \frac{1}{T} \int_{-\frac{T}{2}}^{\frac{T}{2}} \Phi_0^j(x) \exp\left(\frac{2\pi i x \nu}{T}\right) dx$$

$$= \overline{\eta}_\nu \cdot \frac{1}{T} \int_{-\frac{T}{2}}^{\frac{T}{2}} \phi(2^j x) \exp\left(\frac{2\pi i x \nu}{T}\right) dx$$

$$= \overline{\eta}_\nu \frac{1}{2^j T} \int_{-\frac{T}{2}}^{\frac{T}{2}} \phi(y) \exp\left(\frac{2\pi i y \nu}{2^j T}\right) dy \quad \text{(refer to (i))}.$$

Thus

$$\overline{\eta}_\nu \int_{-\frac{T}{2}}^{\frac{T}{2}} \phi(y) \exp(2\pi i y \nu / 2^j T) dy = 0.$$

Letting $j \to \infty$, we have

$$\bar{\eta}_\nu \int_{-\frac{T}{2}}^{\frac{T}{2}} \varphi(y) dy = 0.$$

From (iii), we have $\eta_\nu = 0, \nu \in \mathbf{Z}$. It implies $g(x) \equiv 0$ and $V^\perp = \{0\}$.　□.

Define functions

$$C_\alpha^j(x) = \sum_{\lambda=0}^{K(j)-1} \cos \frac{2\pi \lambda \alpha}{K(j)} \Phi_\lambda^j(x), \quad \alpha \in \mathbf{Z}$$

$$S_\alpha^j(x) = \sum_{\lambda=0}^{K(j)-1} \sin \frac{2\pi \lambda \alpha}{K(j)} \Phi_\lambda^j(x), \quad \alpha \in \mathbf{Z}. \tag{2.4.21}$$

Let S_v^j be the class of functions: $S_v^j = \{C_\alpha^j : \alpha = 0, \cdots, N_j; S_\alpha^j : \alpha = 1, \cdots, N_j - 1\}$ where $N_j = K(j)/2$.

Although the set of functions $\{\Phi_\alpha^j : \alpha = 0, \cdots, K(j)-1\}$ is a basis of V_j, it is not an orthogonal basis. In the following we shall give an orthogonal, real-valued basis for V_j.

Theorem 2.4.2　S_v^j is an orthogonal basis for V_j.

Proof.　Since $\Phi_{\lambda_1}^j(x) = \Phi_0^j(x - \lambda_1 h_j)$, for $\alpha_1 \neq \alpha_2, 0 \leq \alpha_1, \alpha_2 \leq N_j$, we have

$$\langle C_{\alpha_1}^j, C_{\alpha_2}^j \rangle$$

$$= \sum_{\lambda_1=0}^{K(j)-1} \sum_{\lambda_2=0}^{K(j)-1} \cos \frac{2\pi \lambda_1 \alpha_1}{K(j)} \cos \frac{2\pi \lambda_2 \alpha_2}{K(j)} \langle \Phi_0^j(\cdot), \Phi_0^j(\cdot + (\lambda_1 - \lambda_2) h_j) \rangle$$

$$= \sum_{\lambda_1=0}^{K(j)-1} \sum_{\mu=0}^{K(j)-1} \cos \frac{2\pi \lambda_1 \alpha_1}{K(j)} \cos \frac{2\pi (\lambda_1 + \mu) \alpha_2}{K(j)} \langle \Phi_0^j, \Phi_\mu^j \rangle = 0. \tag{2.4.22}$$

In deriving the above expression we have used the following formula heavily

$$\sum_{\lambda=0}^{K(j)-1} \exp\left(\frac{2\pi i \lambda (a_1 - a_2)}{K(j)}\right) = K(j) \delta_{a_1, a_2}, \quad 0 \leq a_1, a_2 \leq K(j) - 1.$$

By a similar approach we can prove:

$$\langle C^j_\alpha, S^j_\beta \rangle = 0, \quad 0 \le \alpha \le N_j, \quad 1 \le \beta \le N_j - 1, \qquad (2.4.23)$$

$$\langle S^j_{\beta_1}, S^j_{\beta_2} \rangle = 0, \quad \beta_1 \ne \beta_2, \quad 1 \le \beta_1, \beta_2 \le N_j - 1. \qquad (2.4.24)$$

From (2.4.22), (2.4.23) and (2.4.24), we know that S^j is an independent system, $\dim S^j = K(j)$, and from (2.4.21) each function in S^j is also in V_j.
□.

In the following theorem we give the two scale equations, which differ from the previous ones in such a way that each of the refinable equations contains 4 terms.

Theorem 2.4.3 Let C^j_α, S^j_α be defined as (2.4.21). Suppose $\varphi(x)$ satisfies the two scale equation (2.4.1). Set

$$\sigma^j_\alpha = \sum_{|\mu| \le p} C_\mu \cos \frac{2\pi\mu\alpha}{K(j)}, \quad \delta^j_\alpha = \sum_{|\mu| \le p} C_\mu \sin \frac{2\pi\mu\alpha}{K(j)}. \qquad (2.4.25)$$

Then, we have the following refinable equations

$$C^j_\alpha(x) = \tfrac{1}{2}\{\sigma^{j+1}_\alpha C^{j+1}_\alpha(x) + \delta^{j+1}_\alpha S^{j+1}_\alpha(x) + \sigma^{j+1}_{K(j)-\alpha} C^{j+1}_{K(j)-\alpha}(x)$$

$$+ \delta^{j+1}_{K(j)-\alpha} S^{j+1}_{K(j)-\alpha}(x)\}, \quad \text{for} \ \ 0 \le \alpha \le N_j$$

$$S^j_\alpha(x) = \tfrac{1}{2}\{-\delta^{j+1}_\alpha C^{j+1}_\alpha + \delta^{j+1}_{K(j)-\alpha} C^{j+1}_{K(j)-\alpha}(x) + \sigma^{j+1}_\alpha S^{j+1}_\alpha(x)$$

$$- \sigma^{j+1}_{K(j)-\alpha} S^{j+1}_{K(j)-\alpha}(x)\} \quad \text{for} \ \ 1 \le \alpha \le N_j - 1. \tag{2.4.26}$$

Proof. In the discussion below, we will use the inverse transform of (2.4.21), i.e.,

$$\Phi^{j+1}_\beta(x)$$

$$= \frac{1}{K(j+1)} \sum_{\alpha=0}^{K(j+1)-1} \left[C^{j+1}_\alpha(x) \cos \frac{2\pi\alpha\beta}{K(j+1)} + S^{j+1}_\alpha(x) \sin \frac{2\pi\alpha\beta}{K(j+1)} \right].$$

$$\tag{2.4.27}$$

For $1 \le \alpha \le N_j - 1$,

$$
\begin{aligned}
C_\alpha^j(x) &= \sum_{\lambda=0}^{K(j)-1} \cos \frac{2\pi\lambda\alpha}{K(j)} \Phi_\lambda^j(x) \\
&= \sum_{\lambda=0}^{K(j)-1} \cos \frac{2\pi\lambda\alpha}{K(j)} \sum_{k=-p}^{p} C_k \Phi_{k+2\lambda}^{j+1}(x) \\
&= \sum_{\lambda=0}^{K(j)-1} \cos \frac{2\pi\lambda\alpha}{K(j)} \sum_{|k|\le p} C_k \frac{1}{K(j+1)} \\
&\quad \times \sum_{\nu=0}^{K(j+1)-1} \left\{ C_\nu^{j+1}(x) \cos \frac{2\pi(k+2\lambda)\nu}{K(j+1)} + S_\nu^{j+1}(x) \sin \frac{2\pi(k+2\lambda)\nu}{K(j+1)} \right\} \\
&= \frac{1}{K(j+1)} \sum_{|k|\le p} C_k \sum_{\nu=0}^{K(j+1)-1} \left[C_\nu^{j+1}(x) \cos \frac{\pi k\nu}{K(j)} + S_\nu^{j+1}(x) \sin \frac{\pi k\nu}{K(j)} \right] \\
&\quad \times \sum_{\lambda=0}^{K(j)-1} \cos \frac{2\pi\lambda\alpha}{K(j)} \cos \frac{2\pi\lambda\nu}{K(j)} \\
&= \frac{1}{K(j+1)} \sum_{\nu=0}^{K(j+1)-1} [C_\nu^{j+1}(z)\sigma_\nu^{j+1} + S_\nu^{j+1}(x)\delta_\nu^{j+1}] I_{\alpha,\nu}.
\end{aligned}
$$

$$(2.4.28)$$

Now we calculate $I_{\alpha,\nu}$ for $1 \le \alpha \le N_j - 1, 0 \le \nu \le K(j+1) - 1$, where

$$
\begin{aligned}
I_{\alpha,\nu} &= \sum_{\lambda=0}^{K(j)-1} \cos \frac{2\pi\lambda\alpha}{K(j)} \cos \frac{2\pi\lambda\nu}{K(j)} \\
&= \frac{1}{2} \sum_{\lambda=0}^{K(j)-1} \left[\cos \frac{2\pi\lambda(\alpha-\nu)}{K(j)} + \cos \frac{2\pi\lambda(\alpha+\nu)}{K(j)} \right],
\end{aligned}
$$

$$(2.4.29)$$

$$
I_{\alpha,\nu} = \begin{cases}
\dfrac{K(j)}{2} \delta_{\nu,\alpha}, & 1 \le \nu \le N_j - 1 \\[2mm]
\dfrac{K(j)}{2} \delta_{\nu,K(j)-\alpha} & N_j \le \nu \le K(j) - 1 \\[2mm]
\dfrac{K(j)}{2} \delta_{\nu,K(j)+\alpha} & K(j) \le \nu \le K(j) + N_j - 1 \\[2mm]
\dfrac{K(j)}{2} \delta_{\nu,K(j+1)-\alpha} & K(j) + N_j \le \nu \le K(j+1) - 1.
\end{cases}
$$

We also have

$$\sigma^j_{K(j)+\alpha} = \sigma^j_{K(j)-\alpha}, \quad \delta^j_{K(j)+\alpha} = -\delta^j_{K(j)-\alpha},$$

$$C^{j+1}_{K(j)+\alpha} = C^{j+1}_{K(j)-\alpha}, \quad S^{j+1}_{K(j)+\alpha} = -S^{j+1}_{K(j)-\alpha}.$$

From (2.4.28) and (2.4.29), for $1 \le \alpha \le N_j - 1$, we have

$$C^j_\alpha(x) = \tfrac{1}{2}\{\sigma^{j+1}_\alpha C^{j+1}_\alpha(x) + \delta^{j+1}_\alpha S^{j+1}_\alpha(x) + \sigma^{j+1}_{K(j)-\alpha}C^{j+1}_{K(j)-\alpha}(x) \tag{2.4.30}$$
$$+\delta^{j+1}_{K(j)-\alpha}S^{j+1}_{K(j)-\alpha}(x)\}.$$

For $\alpha = 0$, along the same approach, we have

$$C^j_0(x) = \frac{1}{2}[\sigma^{j+1}_0 C^{j+1}_0(x) + \sigma^{j+1}_{K(j)}C^{j+1}_{K(j)}(x)]. \tag{2.4.31}$$

For $\alpha = N_j$, we obtain

$$C^j_{N_j}(x) = \sigma^{j+1}_{N_j}C^{j+1}_{N_j}(x) + \delta^{j+1}_{N_j}S^{j+1}_{N_j}(x). \tag{2.4.32}$$

We note that (2.4.30) is (2.4.26), (2.4.31) and (2.4.32) are just the special cases of (2.4.26) since $\delta^{j+1}_0 = \delta^{j+1}_{K(j)} = 0$ and $K(j) - N_j = N_j$. The proof of other equalities are similar. □.

Theorem 2.4.3 establishes the relations between the bases of V_j and V_{j+1}. Now we define W_j as the orthogonal complement of V_j in V_{j+1}, that is $W_j \perp V_j$ and $V_{j+1} = V_j \oplus W_j$. From (2.4.1), (2.4.6) and Theorem 2.4.1 (b), we conclude that

$$W_j \perp W_r \quad \text{for} \quad j \ne r \tag{2.4.33}$$

and

$$\overset{\circ}{L}_2(0,T) = V_0 \oplus \oplus_{j \ge 0}W_j \tag{2.4.34}$$

where $\oplus$ stands for the orthogonal summation.

We construct an orthogonal basis for W_j in the following.

Let $C^j_\alpha(x), S^j_\alpha(x)$ and $\sigma^j_\alpha, \delta^j_\alpha$ be defined as in (2.4.21) and (2.4.25) respectively. Define

$$\tilde{C}^j_\alpha = \frac{C^j_\alpha}{\|C^j_\alpha\|}, \quad \tilde{S}^j_\alpha = \frac{S^j_\alpha}{\|S^j_\alpha\|}, \quad \tilde{\sigma}^j_\alpha = \sigma^j_\alpha \cdot \|C^j_\alpha\|, \quad \tilde{\delta}^j_\alpha = \delta^j_\alpha \cdot \|S^j_\alpha\|,$$

$$A^j_\alpha(x) = \tilde{\sigma}^{j+1}_{K(j)-\alpha}\tilde{C}^{j+1}_\alpha + \tilde{\delta}^{j+1}_{K(j)-\alpha}\tilde{S}^{j+1}_\alpha - \tilde{\sigma}^{j+1}_\alpha \tilde{C}^{j+1}_{K(j)-\alpha} - \tilde{\delta}^{j+1}_\alpha \tilde{S}^{j+1}_{K(j)-\alpha},$$

$$B^j_\alpha(x) = \tilde{\delta}^{j+1}_{K(j)-\alpha}\tilde{C}^{j+1}_\alpha - \tilde{\sigma}^{j+1}_{K(j)-\alpha}\tilde{S}^{j+1}_\alpha + \tilde{\delta}^{j+1}_\alpha \tilde{C}^{j+1}_{K(j)-\alpha} - \tilde{\sigma}^{j+1}_\alpha \tilde{S}^{j+1}_{K(j)-\alpha}$$

$$\tag{2.4.35}$$

and

$$A^j_0(x) = \tilde{\sigma}^{j+1}_{K(j)}\tilde{C}^{j+1}_0 - \tilde{\delta}^{j+1}_0 \tilde{C}^{j+1}_{K(j)}, \quad A^j_{N_j}(x) = 2(\tilde{\delta}^{j+1}_{N_j}\tilde{C}^{j+1}_{N_j} - \tilde{\sigma}^{j+1}_{N_j}\tilde{S}^{j+1}_{N_j}).$$

Theorem 2.4.4

$S^j_w = \{A^j_\alpha : 0 \le \alpha \le N_j; B^j_\alpha : 1 \le \alpha \le N_j - 1\}$ is an orthogonal basis for W_j where A^j_α, and B^j_α are defined in (2.4.35). We call them the real Periodic Quasi-Wavelets (RPQW in abbr.).

Proof. From the definitions given in (2.4.35), we know that each function in S^j_w is in V_{j+1}. A simple calculation shows that

$$\langle A^j_{\alpha_1}, C^j_{\alpha_2}\rangle = \langle A^j_{\alpha_1}, S^j_{\alpha_2}\rangle = \langle B^j_{\alpha_1}, C^j_{\alpha_2}\rangle = \langle B^j_{\alpha_1}, S^j_{\alpha_2}\rangle = 0, \tag{2.4.36}$$

it in turn implies $S^j_w \subset W_j$.

Moreover,

$$\langle A^j_{\alpha_1}, A^j_{\alpha_2}\rangle = 0 \quad \text{for} \quad \alpha_1 \ne \alpha_2, \quad 0 \le \alpha_1, \alpha_2 \le N_j,$$

$$\langle B^j_{\alpha_1}, B^j_{\alpha_2}\rangle = 0 \quad \text{for} \quad \alpha_1 \ne \alpha_2, \quad 1 \le \alpha_1, \alpha_2 \le N_j - 1, \tag{2.4.37}$$

$$\langle A^j_{\alpha_1}, B^j_{\alpha_2}\rangle = 0 \quad \text{for} \quad 0 \le \alpha_1 \le N_j, \quad 1 \le \alpha_2 \le N_j - 1,$$

and

$$\langle A_\alpha^j, A_\alpha^j \rangle = \|\tilde{\sigma}_{K(j)-\alpha}^{j+1}\|^2 + \|\tilde{\delta}_{K(j)-\alpha}^{j+1}\|^2 + \|\tilde{\sigma}_\alpha^{j+1}\|^2 + \|\tilde{\delta}_\alpha^{j+1}\|^2 \neq 0$$

$$\text{for} \quad 0 \leq \alpha \leq N_j$$

$$\langle B_\alpha^j, B_\alpha^j \rangle = \|\tilde{\delta}_{K(j)-\alpha}^{j+1}\|^2 + \|\tilde{\sigma}_{K(j)-\alpha}^{j+1}\|^2 + \|\tilde{\delta}_\alpha^{j+1}\|^2 + \|\tilde{\sigma}_\alpha^{j+1}\|^2 \neq 0$$

$$\text{for} \quad 1 \leq \alpha \leq N_j - 1.$$

$$(2.4.38)$$

It follows that S_w^j is an orthogonal basis for W_j. The proof is now completed. □.

From (2.4.26), we see that the right hand side of each of the equations involves four terms. However, when the underlying function φ is symmetric only two terms are involved in each formula.

Theorem 2.4.5 If $\varphi(-x) = \varphi(x)$, then the formulas for the refinable equations are

$$C_\alpha^j(x) = \tfrac{1}{2}\{\sigma_\alpha^{j+1} C_\alpha^{j+1}(x) + \sigma_{K(j)-\alpha}^{j+1} C_{K(j)-\alpha}^{j+1}(x)\}, \quad \text{for} \quad 0 \leq \alpha \leq N_j,$$

$$(2.4.39)$$

$$S_\alpha^j(x) = \tfrac{1}{2}\{\sigma_\alpha^{j+1} S_\alpha^{j+1}(x) - \sigma_{K(j)-\alpha}^{j+1} S_{K(j)-\alpha}^{j+1}(x)\}, \quad \text{for} \quad 1 \leq \alpha \leq N_j - 1.$$

Proof. By (2.4.1), we have

$$\varphi(-x) = \sum_{|\lambda|\leq p} C_\lambda \varphi(-2x - \lambda h) = \sum_{|\lambda|\leq p} C_{-\lambda}\varphi(2x - \lambda h),$$

but

$$\varphi(-x) = \varphi(x) = \sum_{|\lambda|\leq p} C_\lambda \varphi(2x - \lambda h),$$

thus

$$\sum_{|\lambda|\leq p} (C_\lambda - C_{-\lambda})\varphi(2x - \lambda h) = 0 \quad \text{for} \quad x \in R.$$

It follows that $C_\lambda = C_{-\lambda}$ since $\{\varphi(\cdot - \lambda h)\}_{\lambda=-p}^p$ is an independent system. From (2.4.25), $\delta_\alpha^j = 0$ for all α. Finally, (2.4.39) is obtained directly from (2.4.26). □.

In the following we shall show that the scaling functions converge to cosine and sine functions respectively.

We suppose that $\varphi(x)$ is continuous, $\text{supp}\varphi \subset \left[-\frac{1}{2}T, \frac{1}{2}T\right]$ and satisfies

$$\sum_{k\in\mathbf{Z}} \varphi(x+kh) = 1 \quad \text{for} \quad x \in R. \tag{2.4.40}$$

Define the operator $A^j : C[0,T] \to V_j$ by

$$A^j f(x) = \sum_{\mu=0}^{K(j)-1} f(\mu h_j)\Phi_\mu^j(x), \quad f \in C[0,T]. \tag{2.4.41}$$

For the estimate of error $\|A^j f - f\|_\infty$ we have the following:

Lemma 2.4.1 Under the condition (2.4.40) we have

$$\sum_{\alpha=0}^{K(j)-1} \Phi_\alpha^j(x) = 1, \quad x \in R \tag{2.4.42}$$

and

$$\sup_{x\in[0,T]} |\Phi_\alpha^j(x)| \leq M_\alpha := \sup_{x\in[0,T]} |\varphi(x)|, \tag{2.4.43}$$

for any $\alpha, 0 \leq \alpha \leq K(j) - 1$.

Proof.

$$\sum_{\alpha=0}^{K(j)-1} \Phi_\alpha^j(x) = \sum_{\alpha=0}^{K(j)-1} \sum_{\lambda\in\mathbf{Z}} \varphi(2^j x + \lambda^j T - \alpha h)$$

$$= \sum_{\lambda\in\mathbf{Z}} \sum_{\alpha=0}^{K(j)-1} \varphi(2^j x + \lambda 2^j T - \alpha h)$$

$$(\text{since } \varphi \text{ is compactly supported})$$

$$= \sum_{\lambda\in\mathbf{Z}} \varphi(2^j x + \lambda h) = 1 \quad \text{for any } x, x \in R,$$

hence (2.4.42) is proved.

Set $yh = 2^j xh$. It is clear that $y \in [\nu, \nu + 1), y = \nu + \epsilon,\ 0 \leq \epsilon < 1, 0 \leq \nu \leq K(j) - 1$, and $N = \frac{K}{2}$ is an integer. Since

$$\Phi^j_\alpha(xh) = \sum_{\lambda \in \mathbf{Z}} \varphi(2^j xh + \lambda 2^j Kh - \alpha h),$$

we have:

$$\Phi^j_\alpha(xh) = \sum_{\lambda \in \mathbf{Z}} \varphi(yh + \lambda 2^j Kh - \alpha h). \qquad (2.4.44)$$

From the compactness of the support of φ, there can only be three cases, λ takes $-1, 0, 1$ in (2.4.44):

$$-\frac{K}{2} \leq \nu - 2^j K - \alpha \leq \frac{K}{2} - 1, \quad \lambda = -1$$

$$-\frac{K}{2} \leq \nu - \alpha \leq \frac{K}{2} - 1, \quad \lambda = 0$$

$$-\frac{K}{2} \leq \nu + 2^j K - \alpha \leq \frac{K}{2} - 1, \quad \lambda = 1.$$

but these three cases are mutually exclusive, i.e., only one of the three can be valid. More exactly, when y is fixed, only one term appears in the right side of (2.4.44), hence we obtain (2.4.43). $\square$.

Theorem 2.4.6   Let the operator A^j be defined as in (2.4.41). Then for any $f \in C[0, T]$, we have the estimate

$$\|A^j f - f\|_\infty \leq C\omega\left(f, \frac{K}{2}h_j\right) \qquad (2.4.45)$$

where $C = K \cdot M_\varphi, M_\varphi$ as in (2.4.43) and $\omega(f, \xi)$ is the modulus of continuity.

Proof.   By (2.4.42),

$$|A^j f(x) - f(x)| \leq \sum_{\mu=0}^{K(j)-1} |f(x) - f(\mu h_j)| \cdot |\Phi^j_\mu(x)|.$$

If $\Phi^j_\mu(x) \neq 0$, then there is some λ, which satisfies

$$-\frac{K}{2} \leq \left[\frac{x}{h_j}\right] - \mu + \lambda K(j) \leq \frac{K}{2} - 1. \qquad (2.4.46)$$

Since $[h/xh_j] \in [0, K(j)]$, $\mu \in [0, K(j) - 1]$, we have:

$$1 - K(j) \le \left[\frac{x}{h}\right] - \mu \le K(j). \tag{2.4.47}$$

Hence λ takes only three values $-1, 0, 1$ and (2.4.46) becomes

$$-\frac{K}{2} \le \left[\frac{x}{h_j}\right] - \mu - K(j) \le \frac{K}{2} - 1, \quad \lambda = -1 \tag{2.4.48$_1$}$$

$$-\frac{K}{2} \le \left[\frac{x}{h_j}\right] - \mu \le \frac{K}{2} - 1, \quad \lambda = 0 \tag{2.4.48$_2$}$$

$$-\frac{K}{2} \le \left[\frac{x}{h_j}\right] - \mu + K(j) \le \frac{K}{2} - 1, \quad \lambda = 1. \tag{2.4.48$_3$}$$

The other values of λ (i.e. $|\lambda| \ge 2$) in (2.4.46) contradict (2.4.47).

Since

$$\Phi^j_\mu(x) = \Phi^j_{\mu+K(j)}(x) = \Phi^j_{\mu-K(j)}(x), \quad 0 \le \mu \le K(j) - 1, \tag{2.4.49}$$

for any μ $(0 \le \mu \le K(j)-1)$, by (2.4.49), this μ may be replaced by $\mu + K(j)$ or by $\mu - K(j)$. Let ξ be any one of the three values $\mu, \mu + K(j), \mu - K(j)$. Then from (2.4.48)$_1$–(2.4.48)$_3$ we have

$$-\frac{K}{2} \le \left[\frac{x}{h_j}\right] - \xi \le \frac{K}{2} - 1$$

or equivalently

$$\left|\left[\frac{x}{h_j}\right] - \xi\right| \le \frac{K}{2}. \tag{2.4.50}$$

From above discussion, we have

$$|A^j f(x) - f(x)| \le \sum_{|\mu - [\frac{x}{h_j}]| \le \frac{K}{2}} |f(x) - f(\mu h_j)| \cdot |\Phi^j_\mu(x)|$$

$$\le C\omega\left(f, \frac{K}{2} h_j\right)$$

where $C = K \cdot M_\varphi$. $\square$

Remark $A^j f$ converges uniformly to f on $[0, T]$ as j tends to infinity. In contrast, the set of the functions whose Fourier series do not uniformly converge is dense in $C[0, T]$ (cf. [R]).

Corollary If φ is continuous, supp $\varphi \subset \left[-\frac{1}{2}T, \frac{1}{2}T\right]$ and satisfies (2.4.40) and C_α^j, S_α^j are defined by (2.4.21), then

$$\lim_{j \to \infty} C_\alpha^j(x) = \cos \frac{2\pi\alpha x}{T}, \quad \alpha = 0, 1, \cdots \quad (2.4.51)$$

$$\lim_{j \to \infty} S_\alpha^j(x) = \sin \frac{2\pi\alpha x}{T}, \quad \alpha = 1, 2, \cdots \quad (2.4.52)$$

(2.4.51) and (2.4.52) indicate that the scaling functions $C_\alpha^j(x)$ and $S_\alpha^j(x)$ have stationary properties.

In the following we present some examples.

Example 1 $\varphi := B_0^n$ (refer to (2.1.2)), φ is symmetric about origin when n is odd, and

$$\Phi_\alpha^j(x) := \tilde{B}_{\xi(\alpha)}^{n,j}, \quad \text{where} \quad \xi(\alpha) := \alpha + \hat{n}_0, \hat{n}_0 = \left[\frac{n}{2}\right] + 1,$$

where $\tilde{B}_i^{n,j}$ is defined as in (2.1.12). Then we can construct C_α^j, S_α^j and A_α^j, B_α^j as in (2.4.21) and (2.4.35). These are real periodic polynomial spline quasi-scaling functions and quasi-wavelets, respectively.

Example 2 $\varphi(X) = T_{-n-1,m}\left(x - \frac{1}{2}h\right)$. Let m be even $m = 2n$, $n \geq 0$, $x_\nu = \nu h, h = 2\pi/K, K > m+2$. The basic trigonometric polynomial spline $T_{\alpha,m}(x)$ is defined as

$$T_{\alpha,m}(x) = \begin{cases} \dfrac{1}{\sin \frac{1}{2}h}, & \text{for } x_\alpha \leq x < x_{\alpha+1}, \quad m = 0 \\[2ex] [x_\alpha, \cdots, x_{\alpha+m+1}]_t \left(\sin \frac{t-x}{2}\right)_+^m, & \text{for } x_{\alpha+m+1} > x_\alpha, \quad m > 0 \end{cases} \quad (2.4.53)$$

where

$$(\sin x)_+ = \begin{cases} \sin x, & \text{for } x > 0 \\ 0, & \text{for } x \leq 0. \end{cases}$$

Note $T_{\alpha,m}$ is given by trigonometric divided difference, namely

$$[x_0, \cdots, x_m]f$$

$$= \frac{2^m \det \begin{bmatrix} \cos\frac{x}{2} & \sin\frac{x}{2} & \cdots & \cos\left(n-\frac{1}{2}\right)x & \sin\left(n-\frac{1}{2}\right)x & f(x) \\ x_0 & x_1 & \cdots & x_{m-2} & x_{m-1} & x_m \end{bmatrix}}{\det \begin{bmatrix} 1 & \cos x & \sin x & \cdots & \cos nx & \sin nx \\ x_0 & x_1 & x_2 & \cdots & x_{m-1} & x_m \end{bmatrix}}$$

(2.4.54)

where $m = 2n$ and

$$\det \begin{bmatrix} \varphi_1(x) & \cdots & \varphi_m(x) \\ x_1 & \cdots & x_m \end{bmatrix} := \det \begin{bmatrix} \varphi_1(x_1) & \varphi_2(x_1) & \cdots & \varphi_m(x_1) \\ \varphi_1(x_2) & \varphi_2(x_2) & \cdots & \varphi_m(x_2) \\ & & \cdots & \\ \varphi_1(x_m) & \varphi_2(x_m) & \cdots & \varphi_m(x_m) \end{bmatrix}.$$

The function $T_{\alpha,m}$ has the following properties:

(1) $\mathrm{supp}\, T_{\alpha,m} = [\alpha h, (\alpha + m + 1)h]$,

(2) $T_{\alpha,m} \in C^{m-1}(R)$.

If we use the knots $\{x_\alpha^j\}_{\alpha \in \mathbb{Z}}$, $(x_\alpha^j := x_\alpha/2^j, j \in \mathbb{N})$ instead of $\{x_\alpha\}$, then correspondingly, we can define $T_{\alpha,m}^j$ as $T_{\alpha,m}(= T_{\alpha,m}^0)$ by (2.4.53) and (2.4.54).

Now set

$$\varphi(x) := T_{-n-1,m}\left(x - \frac{1}{2}h\right), x \in [0, 2\pi)$$

We can also construct C_α^j, S_α^j and A_α^j, B_α^j as in (2.4.21) and (2.4.35).

We claim that $\varphi(x)$ is real and even:

$$\varphi(x) = \varphi(-x).$$

(2.4.55)

In fact, this can be directly derived from the Fourier expansion of $\tilde{T}_{j,m}(x) := T_{j,m}(x) + T_{j-K,m}(x)$. From [C9],

$$\tilde{T}_{j,m}(x) = \sum_{\nu \in \mathbb{Z}} b_{j,m}^{(0)}(\nu)e^{i\nu x}, \quad x \in [0, 2\pi]$$

(2.4.56)

where

$$b_{j,m}^{(0)}(\nu) = b_{0,m}^{(0)}(\nu)\omega^{-j\nu}, \qquad \omega = \exp(ih)$$

(2.4.57)

$$b_{0,m}^{(0)}(\nu) = K_m^{(0)} \left[\prod_{k=-n}^{n} C_{\nu+k}^{(0)} \right] \left[\frac{2i\omega^{(m+1)(m+2)/4}}{\omega^{m+1} + 1} \right]$$

(2.4.58)

$$K_m^{(0)} = \frac{m!(2\pi i)^m}{\prod\limits_{\lambda=1}^{m}(\omega^\lambda - 1)}, \quad C_0^{(0)} = \frac{1}{K}, \quad C_j^{(0)} = \frac{1 - \omega^{-j}}{2\pi ij} \qquad (2.4.59)$$

if $j \neq 0$.

From above expressions, we can easily check that

$$b_{-n-1}^{(0)}(-\nu)\omega_1^\nu = b_{-n-1}^{(0)}(\nu)\omega_1^{-\nu}$$

where $\omega_1 = \omega^{\frac{1}{2}}$; that is

$$\tilde{T}_{-n-1,m}\left(x - \frac{h}{2}\right) = \tilde{T}_{-n-1,m}\left(-x - \frac{h}{2}\right). \qquad (2.4.60)$$

Since (2.4.60) is obtained from the Fourier expansion of $\tilde{T}_{-n-1,m}(x)$, its Fourier expansion is valid on the whole real axis, and

$$\tilde{T}_{-n-1,m}\left(x - \frac{h}{2}\right) = \varphi(x), \quad \text{if} \quad x \in [-(n+1)h, (n+1)h] \quad (= \text{supp } \varphi)$$

thus we have (2.4.55).

From Theorem 2.4.5, each of the two scale equations in Example 1 or Example 2 implies the least number of terms: only two terms at the right hand side of each equation, and the decomposition and reconstruction for the coefficients will also be very simple. This property is useful in some applications.

Now, if the generator φ does not satisfy all the conditions listed at the beginning of §2.4 can we still construct a basis of V_j such that the conclusions in Theorem 2.4.1 are still true and the number of terms either in the decomposition or in the reconstruction formulas remains the least?

Let us define a class of functions.

Let $P(\theta)$ be a 2π-periodic, continuously differentiable, real valued function with the Fourier decomposition

$$P(\theta) = \sum_{m \in \mathbb{Z}} p_m e^{im\theta}, \quad p_m > 0 \quad \text{for all} \quad m. \qquad (2.4.61)$$

Any such P will be called a PBF (periodic basis function, see [N,W]).

For the generating function in PBF, we can also establish the corresponding results in previous theorems. (cf. p.117 Note 5).

Theorem 2.4.7 Let P be a PBF and

$$V_j := \text{span}\{\Phi_\nu^j(x)\}_{\nu=0}^{K(j)-1},$$

where

$$\Phi_\nu^j(x) = P(x - \nu h_j), \ K(j) = 2^j K, N_j = \frac{1}{2}K(j), \ h_j = \frac{2\pi}{K(j)}, \ j \in \mathbb{N}.$$

Then

(a) $\dim V_j = K(j)$;

(b) $\overline{U_{j\geq 0}V_j} = \overset{\circ}{L}_2(0, 2\pi)$;

(c) if $S_v^j := \{C_\alpha^j : \alpha = 0, \cdots, N_j, S_\alpha^j : \alpha = 1, \cdots, N_j - 1\}$ where C_α^j and S_α^j are as in (2.4.21), then S_v^j is an orthogonal basis for V_j;

(d) each of the functions $C_\alpha^j(x)$ and $S_\alpha^j(x)$ satisfies the corresponding two scale equation (2.4.39) respectively; and the conclusions in (2.4.35) are valid for $A_\alpha^j(x)$ and $B_\alpha^j(x)$.

Proof. Since the functions $C_\alpha^j(x)$ and $S_\alpha^j(x)$ defined in (2.4.21) are the combinations of functions $\{\Phi_\nu^j(x)\}_{\nu=0}^{K(j)-1}$, thus they all belong to V_j. Using the method in the proof of Theorem 2.4.2, we can prove the conclusion in Theorem 2.4.2, hence (c) and (a) are true.

Since $P(x)$ is real and $p_m > 0$, it follows that $p_m = p_{-m}$, and $P(-x) = P(x)$. Replacing C_μ by p_μ in (2.4.25), we still have $\delta_\alpha^j = 0$ for all α. Hence we have two scale equations (2.4.39) and (2.4.35), each of them involving only two terms at the right hand side of the equation. So (d) is true.

Now we prove (b).

We will prove that $\bigcup_{j\geq 0} V_j$ is dense in $\overset{\circ}{C}([0, 2\pi])$, the space of continuous, periodic functions on $[0, 2\pi]$. If the assertion were false, there would exist a functional $f \in (\overset{\circ}{C}[0, 2\pi])^*$ with $\|f\| = 1$ and $f(g) = 0$ for any $g \in \bigcup_{j\geq 0} V_j$. This is because that the PBF P generates V_j's, we have that $f(P(x - \nu h_j)) = 0$ for all $\nu \in \mathbb{Z}, j \geq 0$. By continuity of both P and its translation, it follows that $f(P(x-y)) = 0, y \in [0, 2\pi]$. Because P is continuously differentiable its Fourier series converges uniformly and absolutely.

Then

$$f\left(\sum_{m\in\mathbb{Z}} p_m e^{im(x-y)}\right) = 0$$

yields

$$\sum_{m\in\mathbb{Z}} p_m e^{-imy} f(e^{imx}) = 0.$$

The series on the left hand side of the second equation is uniformly and absolutely convergent, and therefore defines a function $q(y)$ in $\overset{\circ}{C}[0, 2\pi]$. Since $q(y) \equiv 0$, and $p_m > 0$, this implies that $f(e^{imx}) = 0$ for all $m \in \mathbb{Z}$. Consequently $f = 0$, since $\|f\| = 1$. This is a contradiction.   □.

Let $P, V_j, C_\alpha^j, S_\alpha^j, A_\alpha^j$ and B_α^j be the functions and the space of functions as in Theorem 2.4.7. Set $S_v^j = \{C_\alpha^j : \alpha = 0, \cdots, N_j; S_\alpha^j : \alpha = 1, \cdots, N_j - 1\}$ and $S_w^j = \{A_\alpha^j : \alpha = 0, \cdots, N_j; B_\alpha^j : \alpha = 1, \cdots, N_j - 1\}$. We notice that $V_{j+1} = V_j \oplus W_j, j \geq 0$. Since $S_v^j (S_w^j)$ is an orthogonal basis in V_j (W_j), we can normalize the basis $S_v^j(S_w^j)$ such that the normalized set $\tilde{S}_v^j(\tilde{S}_w^j)$ is an orthonormal basis in V_j (W_j). Following the approach introduced in §2.1 and §2.2 we can get the decomposition and reconstruction formulas, which are similar to (2.2.14) and (2.2.15).

Remark The right hand side of each of the decomposition and reconstruction formulas contains only two terms.

Example 3 $P(x) = P_{2n}(x) = 1 + \sum_{m\neq 0} \dfrac{e^{imx}}{m^{2n}}$, $P_{2n}(x)$ is a polynomial of degree $2n$ in one period. For $n = 1$ and 2, they are

$$P_2(x) = \frac{6 - \pi^2}{6} + \frac{1}{2}(x - \pi)^2,$$

$$P_4(x) = \frac{360 - 7\pi^4}{360} + \frac{\pi^2(x - \pi)^2}{12} - \frac{(x - \pi)^4}{24}.$$

§2.5   Other Methods in Periodic Multi-resolution Analysis

In the previous sections, we have introduced various ways to decompose the space $\overset{\circ}{L}_2\,(0, 2\pi)$ into the following sequence of spaces

$$\overset{\circ}{L}_2\,([0, 2\pi]) = V_0 \oplus W_0 \oplus W_1 \oplus \cdots \qquad (2.5.1)$$

where $V_0 \perp W_0$ and $W_k \perp W_l(k \neq l)$.

Besides, they have to comply with the following conditions

$$V_j \subset V_{j+1}, \quad j \in \mathbb{N}, \qquad (2.5.2)$$

$$\overline{\bigcup_{j \geq 0} V_j} = \overset{\circ}{L}_2\,([0, 2\pi]) \qquad (2.5.3)$$

and for $j = 0, 1, \cdots$, there exists $\phi_j \in V_j$ such that

$$V_j = \mathrm{span}\{\phi_j(\cdot), \phi_j(\cdot - h_j), \cdots, \phi_j(\cdot - (K(j) - 1)h_j)\}. \qquad (2.5.4)$$

A sequence of spaces $\{V_j\}_{j \geq 0}$ is called a **periodic multi-resolution (PMR)** if it satisfies (2.5.2)–(2.5.4).

The function ϕ_j in (2.5.4) is called the scaling function.

In §2.1–§2.4, we have constructed different kinds of orthogonal or orthonormal bases. In some occasion, we need bi-orthogonal basis instead.

We also notice that in the non-periodic case, V_j are spanned by shifts of one scaling function for all j. But in the periodic case the scaling function ϕ_j (sometimes we write Φ_0^j, see (2.4.4)) in V_j may be different for different j (see (2.5.4)). On the other hand, from many examples listed in §2.1, §2.2 and §2.4, we know that for the construction of PMR, the first thing is to periodize a function to obtain Φ_0^j; the space V_j is spanned by $\{\Phi_\alpha^j\}_{\alpha=0}^{K(j)-1}$ the h_j-shifts of Φ_0^j.

In the following we shall present a method from which we can construct a PMR by a set of numbers (cf. [KLT]).

Let $S(\nu)$ be the class of all complex periodic sequences of period ν. The space $S(\nu)$ is equipped with the norm

$$\|\alpha\| := \left(\sum_{k=0}^{\nu} |\alpha(k)|^2\right)^{\frac{1}{2}}, \quad \alpha \in S(\nu).$$

Theorem 2.5.1   For $j > 0$, let $f_{j+1} \in S(K(j+1))$ and satisfy

$$1 - f_j(n) = O(j^{-1-\varepsilon}) \quad \text{as} \quad j \to \infty$$

$$|f_{j+1}(k)|^2 + |f_{j+1}(k+K(j))|^2 \le 1, \quad k = 0, \cdots, K(j) - 1$$

and for each $k = 0, \cdots, K(j) - 1$

$$\prod_{l=j+1}^{\infty} |f_l(k + pK(j))|^2 > 0, \quad \text{for some} \quad p \in \mathbf{Z}.$$

Set

$$q_j(n) := \prod_{l=j+1}^{\infty} f_l(n)$$

and

$$\varphi_{j,k}(x) := \sum_{p \in \mathbf{Z}} q_j(k + pK(j)) e^{i(k+pK(j))x}$$

$k = 0, \cdots, K(j) - 1$, and

$$V_j := \text{span}\{\varphi_{j,k} : k = 0, \cdots, K(j) - 1\}.$$

Then $\{\varphi_{j,k}, k = 0, \cdots, K(j) - 1\}$ is an orthogonal basis of V_j, $\{V_j, j \ge 0\}$ forms a PMR of $\overset{\circ}{L}_2([0, 2\pi])$ and $\varphi_j(x) := \sum_{k=0}^{K(j)-1} \varphi_{j,k}(x)$ is the scaling function of V_j.

Let W_j be the orthogonal complement of V_j in V_{j+1}.

Theorem 2.5.2   Let $h_{j+1} \in S(K(j+1))$ and satisfy

$$h_{j+1}(l + K(j)) = -h_{j+1}(l) \ne 0, \quad l = 0, \cdots, K(j) - 1.$$

Define $g_{j+1} \in S(K(j+1))$ by

$$g_{j+1}(l) = h_{j+1}(l) \frac{\overline{f_{j+1}(l + K(j))}}{\|\varphi_{j+1,l}\|^2}, \quad l = 0, \cdots, K(j+1) - 1$$

and

$$\psi_{j,k}(x) = \sum_{p \in \mathbf{Z}} \gamma_j(k + pK(j)) e^{i(k+pK(j))x}$$

where $\gamma_j(k + pK(j)) = g_{j+1}(k)q_{j+1}(k + pK(j + 1)), p \in \mathbf{Z}$. Then $\{\psi_{j,k}, k = 0, \cdots, K(j) - 1\}$ is an orthogonal basis of W_j.

In §2.1–§2.4, we used different kinds of functions to produce φ_j and then defined the subspace V_j which is the linear span of the shifts of the function φ_j. We gave different approaches to prove that the following functions are linearly independent

$$B(\varphi_j) := \{\varphi_j(x), \varphi_j(x - h_j), \cdots, \varphi(x - (K(j) - 1)h_j)\} \qquad (2.5.5)$$

We would now like to ask: are there any general rules to judge whether $B(\varphi_j)$ is the basis of the space:

$$V_j := \mathrm{span}\{\varphi_j(\cdot), \varphi_j(\cdot - h_j), \cdots, \varphi(\cdot - (K(j) - 1)h_j)\}. \qquad (2.5.6)$$

In the previous sections we also gave some different methods of verifying the union of the subspaces $\{V_j\}_{j \geq 0}$ is dense in $\overset{\circ}{L}_2([0, 2\pi])$:

$$\overline{\bigcup_{j \geq 0} V_j} = \overset{\circ}{L}_2([0, 2\pi]). \qquad (2.5.7)$$

Now the second problem arises: are there any criteria for characterizing the above fact? (cf. *p.118* Note 6).

Let $a := (a_u)_{u \in \mathbf{Z}}, b := (b_u)_{u \in \mathbf{Z}}$ be two elements in l^2. The bracket product of level j ($j \in \mathbf{N}$) for a and b is defined by

$$[a, b]_j := ([a, b]_{j,k})_{k=0}^{K(j)-1},$$

where

$$[a, b]_{j,k} := \sum_{\nu \in \mathbf{Z}} a_{k+\nu K(j)} \overline{b_{k+\nu K(j)}}, \quad k = 0, \cdots, K(j) - 1. \qquad (2.5.8)$$

For $f \in \overset{\circ}{L}_2(0, 2\pi)$, denote $C(f) = \{C_u(f)\}_{u \in \mathbf{Z}} \in l^2$, where $C_u(f)$ are the Fourier coefficients

$$C_u(f) = \langle f, e^{iu \cdot} \rangle = \frac{1}{2\pi} \int_0^{2\pi} f(x)e^{-iux} dx. \qquad (2.5.9)$$

It is evident that

$$C_u(f(\cdot - kh_j)) = \omega_j^{uk} C_u(f) \qquad (2.5.10)$$

with $\omega_j = \exp(-2\pi i/K(j))$.

The discrete Fourier transform (DFT) of $(a_k)_{k=0}^{K(j)-1}$ (abbreviated as $DFT(K(j))$, of length $K(j)$ is denoted as $(\hat{a}_\nu)_{\nu=0}^{K(j)-1}$,

$$\hat{a}_\nu = \hat{a}(\nu) = \sum_{k=0}^{K(j)-1} \omega_j^{k\nu} a_k, \quad \nu = 0,\cdots,K(j)-1. \tag{2.5.11}$$

Then the inner product of two functions f and g in $\overset{\circ}{L}_2\ ([0,2\pi])$ can be written into the DFT of the bracket product:

$$\langle f(\cdot - lh_j), g \rangle = \sum_{k=0}^{K(j)-1} [C(f), C(g)]_{j,k} \omega_j^{kl}, \tag{2.5.12}$$
$$= [C(f), \hat{C}(g)]_j(l), \quad l = 0,\cdots,K(j)-1.$$

For $\varphi_j \in \overset{\circ}{L}_2\ ([0,2\pi])$, we consider the system $B(\varphi_j)$ (refer to (2.5.5)). By (2.5.12), the inner product

$$\langle \varphi_j(\cdot - lh_j), \varphi_j(\cdot - kh_j)\rangle = \sum_{\nu=0}^{K(j)-1} \omega_j^{\nu(l-k)}[C(\varphi_j), C(\varphi_j)]_{j,\nu}. \tag{2.5.13}$$

Then the corresponding Gramian matrix is

$$(\langle \varphi_j(\cdot - lh_j), \varphi_j(\cdot - \nu h_j)\rangle)_{l,\nu=0}^{K(j)-1} = G_j(\mathrm{diag}[C(\varphi_j), C(\varphi_j)]_j)\overline{G}_j \tag{2.5.14}$$

where $G_j = (\omega_j^{kl})_{k,l=0}^{K(j)-1}$ is a $K(j) \times K(j)$ matrix.

The following theorem answers, in some extent, the first problem.

Theorem 2.5.3   Let $\varphi_j \in \overset{\circ}{L}_2\ ([0,2\pi])$ and $j \in \mathbb{N}$ be given. Then
(i) $B(\varphi_j)$ is a basis of V_j if and only if

$$[C(\varphi), C(\varphi)]_{j,k} > 0, \quad k = 0,\cdots,K(j)-1, \tag{2.5.15}$$

(ii) $B(\varphi_j)$ is an orthonormal basis of V_j if and only if

$$K(j)[C(\varphi), C(\varphi)]_{j,k} = 1, \quad k = 0\cdots,K(j)-1. \tag{2.5.16}$$

Proof.   (i) $B(\varphi_j)$ is a basis of V_j if and only if the Gramian matrix is non-singular. From (2.5.13), this is equivalent to (2.5.15) to be true.

(ii) $B(\varphi_j)$ is an orthonormal basis of V_j if and only if the Gramian matrix of $B(\varphi_j)$ is equal to I_j, the $K(j)$-th identity matrix. Since $G_j\overline{G}_j = K(j)I_j$, from (2.5.14), $\mathrm{diag}[C(\varphi_j), C(\varphi_j)] = (\overline{G}_jG_j)^{-1} = (K(j)I_j)^{-1} = \frac{1}{K(j)}I_j$, hence (2.5.16).

$\square$.

Corollary Let $\varphi \in \overset{\circ}{L}_2([0,2\pi])$ be a function satisfying (2.5.15). Define a function $\tilde{\varphi}$ via the Fourier coefficients such that

$$C_v(\tilde{\varphi}) := K(j)^{-1/2}[C(\varphi), C(\varphi)]_{j,\tilde{v}}^{-1/2} C_v(\varphi), \qquad (2.5.17)$$

where $\tilde{v} = v \bmod K(j)$.

Then $B(\tilde{\varphi}) := \{\tilde{\varphi}(\cdot), \tilde{\varphi}(\cdot - h_j), \cdots, \tilde{\varphi}(\cdot - (K(j)-1)h_j)\}$ is an orthonormal basis of $V(\varphi) = \mathrm{span}\{\varphi(\cdot), \cdots, \varphi(\cdot - (K(j)-1)h_j\}$.

Proof. Since $\tilde{v}_j = k + lK(j) \bmod K(j) = k \bmod K(j)$.

$$
\begin{aligned}
[C(\tilde{\varphi}), C(\tilde{\varphi})]_{j,k} &= \sum_{l\in\mathbb{Z}} |C_{k+lK(j)}(\tilde{\varphi})|^2 \\
&= K(j)^{-1}[C(\varphi), C(\varphi)]_{j,\tilde{v}}^{-1} \sum_{l\in\mathbb{Z}} |C_{k+lK(j)}(\varphi)|^2 \\
&= K(j)^{-1}.
\end{aligned}
$$

From (2.5.16) and (ii) of Theorem 2.5.3, $B(\tilde{\varphi})$ is an orthonormal basis of $V(\tilde{\varphi}) = \mathrm{span}\{\tilde{\varphi}(\cdot), \cdots, \tilde{\varphi}(\cdot - (K(j-1)h_j)\}$. It remains to prove that $V(\tilde{\varphi}) = V(\varphi)$.

Set

$$\hat{a}_{j,\tilde{v}} = K(j)^{-1/2}[C(\varphi), C(\varphi)]_{j,\tilde{v}}^{-1/2}.$$

There are constants $\{a_{j,l}\}_{l=0}^{K(j)-1}$ such that

$$\hat{a}_{j,l} = \sum_{k=0}^{K(j)-1} a_{j,k} e^{-ilkh_j}. \qquad (2.5.18)$$

In fact, such $a_{j,k}$ may be obtained from

$$a_{j,k} = \frac{1}{K(j)} \sum_{v=0}^{K(j)-1} \hat{a}_{j,v} e^{ikvh_j}. \qquad (2.5.19)$$

Moreover,

$$a_{j,k+K(j)} = a_{j,k}; \quad \hat{a}_{j,k+K(j)} = \hat{a}_{j,k}. \tag{2.5.20}$$

From (2.5.17) and (2.5.18), we have

$$C_v(\tilde{\varphi}) = \hat{a}_{j,v} C_v(\varphi)$$

$$= \sum_{k=0}^{K(j)-1} a_{j,k} e^{-ivkh_j} C_v(\varphi) \tag{2.5.21}$$

$$= C_v \left(\sum_{k=0}^{K(j)-1} a_{j,k} \varphi(\cdot - kh_j) \right), \quad \forall v \in \mathbb{Z}.$$

Hence

$$\tilde{\varphi}(x) = \sum_{k=0}^{K(j)-1} a_{j,k} \varphi(x - kh_j)$$

thus

$$\tilde{\varphi} \in V(\varphi), V(\tilde{\varphi}) \subseteq V(\varphi). \tag{2.5.22}$$

On the other hand, we define

$$\hat{b}_{j,k} := \begin{cases} (\hat{a}_{j,k})^{-1}, & \text{if } a_{j,k} \neq 0 \\ 0, & \text{otherwise.} \end{cases}$$

Then $\hat{b}_{j,k} \in \mathbb{C}$, $\hat{b}_{j,k+K(j)} = \hat{b}_{j,k}$. From (2.5.21), we have $C_v(\varphi) = \hat{b}_{j,v} C_v(\tilde{\varphi})$, and then $\varphi \in V(\tilde{\varphi})$, $V(\varphi) \subseteq V(\tilde{\varphi})$. Considering (2.5.22), we have $V(\tilde{\varphi}) = V(\varphi)$, and therefore $B(\tilde{\varphi})$ is an orthonormal basis of $V(\varphi)$. $\qquad \square$

In the following we characterize the second basic problem (2.5.7).

Theorem 2.5.4 Let $\{V_j\}_{j=0}^{\infty}$ be a nested sequence of V_j which are h_j-shift invariant subspaces (see (2.5.2) and (2.5.4)) with $\varphi_j \in \overset{\circ}{L}_2([0, 2\pi])$. Then we have

$$\overline{\bigcup_{j \geq 0} V_j} = \overset{\circ}{L}_2([0, 2\pi]) \tag{2.5.23}$$

if and only if

$$\bigcup_{j \geq 0} \text{supp } C(\varphi_j) = \mathbb{Z} \tag{2.5.24}$$

where $\mathrm{supp}(f) := \{n \in \mathbb{Z} : C_n(f) \neq 0\}$ of $C(f)$.

Proof. Step 1. Suppose that (2.5.24) is not satisfied. Then there exists

$$n_0 \in \mathbb{Z} \setminus \bigcup_{j \geq 0} \mathrm{supp} C(\varphi_j).$$

Hence

$$C_{n_0}(\varphi_j) = \frac{1}{2\pi} \int_0^{2\pi} \varphi_j(x) e^{-in_0 x} dx = 0$$

for all $j \in \mathbb{N}$, such that $e^{in_0 x} \perp \overline{(\bigcup_{j \geq 0} V_j)}$, thus (2.5.23) is not satisfied.

Step 2. Since $\varphi_j \in V_j \subset V_{j+1}$ there are constants $\{a_{j,n}\}_{n=0}^{K(j+1)-1}$ such that

$$\varphi_j(x) = \sum_{n=0}^{K(j+1)-1} a_{j,n} \varphi_{j+1}(x - nh_{j+1}).$$

Then

$$C_k(\varphi_j) = \left(\sum_{n=0}^{K(j+1)-1} a_{j,n} \omega_{j+1}^{nk} \right) C_k(\varphi_{j+1}),$$

hence

$$\mathrm{supp} C(\varphi_j) \subseteq \mathrm{supp}(C(\varphi_{j+1})) \quad (j \in \mathbb{N}). \tag{2.5.25}$$

Suppose that there exists $f \in \overset{\circ}{L}_2([0, 2\pi])$ with $f \neq 0$ and

$$f \perp \overline{\left(\bigcup_{j \geq 0} V_j \right)}. \tag{2.5.26}$$

We denote an index with the property

$$|C_{k_0}(f)| = \max\{|C_k(f)| : k \in \mathbb{Z}\} > 0 \tag{2.5.27}$$

by $k_0 \in \mathbb{Z}$.

By (2.5.24) and (2.5.25), we conclude that there is an index $j_0 \in \mathbb{N}$ such that $k_0 \in \mathrm{supp}\, C(\varphi_{j_0})$ and $K(j_0) \geq |k_0| + 1$. Since $\varphi_{j_0} \in V_{j+1}$ for all $j \geq j_0$, it follows that $f \perp \tilde{V}_{j+1}(\varphi_{j_0})(j \geq j_0)$, where

$$\tilde{V}_{j+1}(\varphi_{j_0}) := \mathrm{span}\{\varphi_{j_0}(\cdot), \varphi_{j_0}(\cdot - h_{j+1}), \cdots, (K(j+1)-1)h_{j+1}\}.$$

From

$$\langle f(\cdot), \varphi_{j_0}(\cdot - lh_{j+1}) \rangle = 0,$$

we obtain

$$C_{k_0}(f)\overline{C_{k_0}(\varphi_{j_0})} + \sum_{n\in\mathbf{Z},n\neq 0} C_{k_0+nK(j+1)}(f)\overline{C_{k_0+nK(j+1)}(\varphi_{j_0})} = 0, \quad (j \geq j_0).$$

$$(2.5.28)$$

Since

$$\sum_{n\in\mathbf{Z}} |C_n(f)\overline{C_n(\varphi_{j_0})}| \leq \|C(f)\|_{l^2}\|C(\varphi_{j_0})\|_{l^2}$$

$$< \infty,$$

for given $\varepsilon = |C_{k_0}(f)\overline{C_{k_0}(\varphi_{j_0})}| > 0$ we can choose $j_1 \geq j_0$ such that

$$\sum_{|n|\leq K(j_1)} |C_n(f)\overline{C_n(\varphi_{j_0})}| < \frac{\varepsilon}{2}. \qquad (2.5.29)$$

But (2.5.29) contradicts (2.5.28) for $j = j_1$. This implies $f = 0$, i.e., (2.5.23) is satisfied. □.

Theorem 2.5.4 establishes the equivalence between (2.5.23) and (2.5.24) under the conditions (2.5.2) and (2.5.4). Now if we strengthen the conditions to that the class of the h_j-shifts of φ_j is an orthonormal basis of V_j, what is the new characterization for (2.5.23)? The following theorem gives an answer to this problem (cf. p.118, Note 7).

Theorem 2.5.5   Let $\{V_j\}_{j=0}^{\infty}$ be a nested sequence of the h_j-shift invariant subspaces V_j. The class of the h_j-shifts of φ_j, $\{\varphi_j(\cdot - lh_j)\}_{l=0}^{K(j)-1}$, is an orthonormal basis of V_j. Let $\tilde{\varphi}_j = K(j)\varphi_j$. Then we have

$$\overline{\bigcup_{j\geq 0} V_j} = \overset{\circ}{L}_2 \left([0, 2\pi]\right) \qquad (2.5.30)$$

if and only if

$$\lim_{j\to\infty} |C_n(\tilde{\varphi}_j)| = 1, \quad \text{for any} \quad n \in \mathbf{Z}, \qquad (2.5.31)$$

where $C_n(\tilde{\varphi}_j)$ is the nth Fourier coefficient of $\tilde{\varphi}_j$.

Proof. 1. Assume that $\lim_{j\to\infty} |C_n(\tilde{\varphi}_j)| = 1$ for $n \in \mathbf{Z}$. By Theorem 2.5.4, it is sufficient to show that for $n \in \mathbf{Z}$, there exists a $j \in \mathbf{N}$ such that $C_n(\varphi_j) \neq 0$. In fact, by (2.5.31), for any $n \in \mathbf{Z}$ we have $j = j(n), j \in \mathbf{Z}$

such that $C_n(\tilde{\varphi}_j) \neq 0$. Thus there exists a $j \in \mathbf{Z}$, so that $C_n(\varphi_j) \neq 0$. 2. Suppose that (2.5.30) holds.

Let $P_j f$ be the orthogonal projection of $f (\in \overset{\circ}{L}_2 ([0, 2\pi]))$ on V_j. Then the nested property of $\{V_j\}$ and (2.5.30) imply that

$$\lim_{j \to \infty} \|P_j f - f\| = 0, \quad \text{for all} \quad f \in \overset{\circ}{L}_2 ([0, 2\pi]). \tag{2.5.32}$$

In particular, (2.5.32) holds for $f(x) = e^{imx}$. Given $\varepsilon > 0$, there is a $j_0 \in \mathbf{Z}$ such that $\|P_j f - f\| < \varepsilon$ for all $j \geq j_0$. Consequently, we have

$$\begin{aligned}1 = \|f\| &\leq \|P_j f\| + \|f - P_j f\| \\ &< \|P_j f\| + \varepsilon,\end{aligned}$$

hence

$$\|P_j f\| > 1 - \varepsilon \quad \text{for all} \quad j \geq j_0. \tag{2.5.33}$$

By assumption, $\{\varphi_j(\cdot - lh_j)\}_{l=0}^{K(j)-1}$ is an orthonormal basis of V_j, and

$$P_j f(x) = \sum_{k=0}^{K(j)-1} \langle f, \phi_j(\cdot - kh_j)\rangle \phi_j(x - kh_j),$$

from which it follows

$$\begin{aligned}\|P_j f\|^2 &= \sum_{k=0}^{K(j)-1} |\langle f, \phi_j(\cdot - kh_j)\rangle|^2 \\ &= \sum_{k=0}^{K(j)-1} \left|\sum_{n \in \mathbf{Z}} C_n(f)\overline{C_n(\phi_j)}\bar{\omega}_j^{nk}\right|^2 \\ &= \sum_{k=0}^{K(j)-1} \left|\sum_{n \in \mathbf{Z}} \delta_{n,m}\overline{C_n(\phi_j)}\bar{\omega}_j^{nk}\right|^2 \\ &= \sum_{k=0}^{K(j)-1} |C_m(\phi_j)|^2 = K(j)|C_m(\phi_j)|^2 \\ &= |C_m(\tilde{\phi}_j)|^2.\end{aligned}$$

Therefore (by (2.5.33))

$$\lim_{j \to \infty} |C_m(\tilde{\varphi}_j)| \geq 1. \tag{2.5.34}$$

From (2.5.16),

$$K(j) \sum_{l \in \mathbb{Z}} |C_{k+lK(j)}(\varphi_j)|^2 = 1,$$

that is

$$\sum_{l \in \mathbb{Z}} |C_{k+lK(j)}(\tilde{\varphi}_j)|^2 = 1,$$

for any $j \geq 0$. Thus

$$\varlimsup_{j \to \infty} |C_m(\tilde{\varphi}_j)| \leq 1. \tag{2.5.35}$$

now (2.5.34) and (2.5.35) imply (2.5.31).   □.

Note 1   The work [KTM] by Kamada, Toraichi and Mori on how to orthonormalize a periodic spline function was introduced by Professor C. K. Chui to the author in 1990 in Hongchow. We then applied the idea of multi-resolution and the theory of spline functions to develop different kinds of spline scaling functions and quasi-wavelets. Further, we have constructed two kinds of wavelets with different methods; the one presented here is simpler, it can be used to construct a large class of spline quasi-wavelets. The first work on the construction of orthonormal quasi-wavelet was completed at the end of 1992, and appeared in [C10].

Based on properties of periodic shift-invariant spaces and related products, Plonka and Tasche [PT] gave an approach to p-periodic wavelets for general periodic scaling functions. A new method for constructing the periodic wavelets was given in [KLT], which will be introduced in §2.5.

Note 2   The first half of §2.2 is extracted from Chen's work [C2]. The latter half of §2.2 originates from [C4]. Notice that the result in [C4] is weaker; we strengthen it here.

Note 3   This is the extension of the idea of periodic multi-resolution, the contents of this passage is taken from [C8].

Note 4   The main parts of §2.4 are extracted from [CLPX].

Note 5   The definition of PBF was first presented in [NW]. The proof of (b) in Theorem 2.4.7 is quoted from [NW], the proof of (a), (c) and (d) are analogous to that in [CLPX].

Note 6   In §2.5, the Theorem 2.5.1 and Theorem 2.5.2 can be found in [KLT]. We use the bracket product to prove Theorem 2.5.3 and Theorem 2.5.4 which are extracted from [PT].

Note 7   Theorem 2.5.5 is quoted from [LP], the result (2.5.31) is better than (2.5.24), since for some $\tilde{\varphi}_j$, (2.5.31) is easier to check.

Chapter 3

THE APPLICATION OF QUASI-WAVELETS
IN SOLVING A BOUNDARY INTEGRAL
EQUATION OF THE SECOND KIND

Boundary value problems for the two-dimensional Helmholtz equation are usually solved by the boundary integral method, which leads to a Fredholm integral equation of the second kind (see [Ke], [GW], [KS1], [KS2], [Ya1], [Ya2], [Kr])

$$u(x) = \int_0^{2\pi} u(y) \left[a(x-y)\ln \left| 2\sin\frac{x-y}{2} \right| + b(x,y) \right] dy + g(x) \qquad (3.0)_1$$

$x \in [0, 2\pi]$, where

$$a(t) = a_0 + a_1(t)\sin^2 \frac{t}{2} \qquad (3.0)_2$$

a_0 is a constant, $b(x,y)$ is a continuous function of x and y, with period 2π in each variable, $a_1(x)$ and $g(x)$ are continuous periodic functions.

The construction of a conformal mapping can also lead to this kind of integral equation; compared to two decades ago conformal mapping can solve a much wider class of problems (see $p.159$, Note 1). Recently Beylkin and Brewster introduced a method called multi-scale strategy in [BB]. But when we use this method to solve the integral equation $(3.0)_1$ non-diagonal matrices will appear if we use other kinds of wavelets. We therefore appeal to PQW. The advantage of using PQW is that some of the matrices become diagonal (see Theorem 3.2.1), An $O(N^2)$ algorithm is introduced in [CP1]. We want to speed up the algorithm in [CP1] as much as possible. In the following, combining the multi-scale strategy method with PQW, we present a new algorithm with a complexity of only $O(N)$. (cf. $p.160$, Note 2, and $p.163$, Note 10).

In the latter part of this chapter we shall show that the Dirichlet problem also can be solved by this method, hence some related problems in mathematics and physics can be solved by using this method too.

In Chapter 2, §2.1, we have decomposed the space of periodic square integrable functions into hierarchical sub-spaces $\{V_j\}_{j\geq 0}$ such that

$$\mathring{L}_2\left([0, 2\pi]\right) = \overline{\bigcup_{j\geq 0} V_j}, \tag{3.0}_3$$

where

$$V_{j+1} = V_j \oplus W_j \quad \text{and} \quad V_j \subseteq V_{j+1} \tag{3.0}_4$$

We have found the orthonormal bases $\{A_k^{n,m}\}_{k=0}^{K(m)-1}$ and $\{D_k^{n,m}\}_{k=0}^{K(m)-1}$ for the space V_m and the space W_m, respectively.

§3.1   Discretization

We rewrite the integral $(3.0)_1$ in the operator form

$$u = Tu + g. \tag{3.1.1}$$

Let P_m be the projective operator of $\mathring{L}_2\left([0, 2\pi]\right)$ onto V_m. Then the following equation is an approximate version of $(3.0)_1$:

$$u_m = P_m T u_m + P_m g \tag{3.1.2}$$

where $u_m \in V_m$. Suppose that

$$u_m = \sum_{j=0}^{K(m)-1} S_j^m A_j^{n,m}, \quad P_m g = \sum_{j=0}^{K(m)-1} g_j^m A_j^{n,m}. \tag{3.1.3}$$

Substituting (3.1.3) into (3.1.2), we obtain

$$S_j^m = \sum_{k=0}^{K(m)-1} \beta_{j,k} S_k^m + g_j^m, \quad 0 \leq j \leq K(m) - 1 \tag{3.1.4}$$

where

$$\beta_{jk}^m = \langle T A_k^{n,m}, A_j^{n,m} \rangle, \quad S_j^m = \langle u_m, A_j^{n,m} \rangle. \tag{3.1.5}$$

We now have to solve linear system (3.1.4).

Denote

$$\theta(x-y) = a(x-y)\ln\left|2\sin\frac{x-y}{2}\right|. \tag{3.1.6}$$

Then

$$\beta_{jk}^m = e_{jk}^m + f_{jk}^m, \quad 0 \le j \le K(m) - 1 \tag{3.1.7}$$

where

$$e_{jk}^m = \frac{1}{2\pi}\int_0^{2\pi}\int_0^{2\pi}\theta(x-y)A_k^{n,m}(y)\overline{A_j^{n,m}(x)}\,dx\,dy \tag{3.1.8}$$

$$f_{jk}^m = \frac{1}{2\pi}\int_0^{2\pi}\int_0^{2\pi}b(x,y)A_k^{n,m}(y)\overline{A_j^{n,m}(x)}\,dx\,dy \tag{3.1.9}$$

for $0 \le j \le K(m) - 1$, $0 \le k \le K(m) - 1$.

Let E^m and F^m be matrices of size $K(m) \times K(m)$, and let S^m and g^m be column vectors of length $K(m)$, i.e.,

$$E^m = (e_{jk}^m), \quad F^m = (f_{jk}^m), \quad S^m = (S_j^m), \quad g^m = (g_j^m) \tag{3.1.10}$$

then (3.1.4) is equivalent to

$$S^m = (E^m + F^m)S^m + g^m. \tag{3.1.11}$$

Let M_m be the matrix defined in (2.1.43). Multiply M_m to both sides of (3.9) and split into a pair of equations to obtain

$$S_{m-1}^m = (E_{ss}^{m-1} + F_{ss}^{m-1})S_{m-1}^m + (E_{sd}^{m-1} + F_{sd}^{m-1})d_{m-1}^m + g_s^{m-1} \tag{3.1.12}$$

$$d_{m-1}^m = (E_{ds}^{m-1} + F_{ds}^{m-1})S_{m-1}^m + (E_{dd}^{m-1} + F_{dd}^{m-1})d_{m-1}^m + g_d^{m-1} \tag{3.1.13}$$

where

$$S_{m-1}^m = L_m S^m, \quad d_{m-1}^m = H_m S^m, \quad g_s^{m-1} = L_m g^m, \quad g_d^{m-1} = H_m g^m \tag{3.1.14}$$

$$E_{ss}^{m-1} = L_m E^m L_m^T, \quad E_{sd}^{m-1} = L_m E^m H_m^T, \quad E_{ds}^{m-1} = H_m E^m L_m^T \tag{3.1.15}$$

$$E_{dd}^{m-1} = H_m E^m H_m^T, \quad F_{ss}^{m-1} = L_m F^m L_m^T, \quad F_{sd}^{m-1} = L_m F^m H_m^T \tag{3.1.16}$$

$$F_{ds}^{m-1} = H_m F^m L_m^T, \quad F_{dd}^{m-1} = H_m F^m H_m^T. \tag{3.1.17}$$

In fact, since $L_m^T L_m + H_m^T H_m = I_m$, from (3.1.11)

$$L_m S^m = L_m (E^m + F^m) S^m + L_m g^m$$

$$= L_m (E^m + E^m) L_m^T L_m S^m + L_m (E^m + F^m) H_m^T H_m S^m + L_m g^m,$$
$$(3.1.18)$$

and

$$H_m S^m = H_m (E^m + F^m) L_m^T L_m S^m + H_m (E^m + F^m) H_m^T H_m S^m + H_m g^m,$$
$$(3.1.19)$$

(3.1.18) and (3.1.19) are just (3.1.12) and (3.1.13) respectively.

§3.2   Simplifying the Procedure by Using PQW

In solving S_{m-1}^m from (3.1.12) and (3.1.13), it involves a great deal of calculations.

There appear a lot of matrices in (3.1.12) and (3.1.13): $E_\&^{m-1}$ and $F_\&^{m-1}$ where $\&$ stands for ss, sd, ds or dd. Even if we use Daubechies' wavelets, these matrices still puzzle us since they are not diagonal.

We therefore appeal to the PQW, it makes the matrices E_q diagonal and the norm of $F_\&$ small. Thus we can neglect $F_\&$ in some occasions, without effecting the order of the error. In turn, this will greatly simplify the procedure in the whole computation.

In the following we shall show that the matrix $E_\&$ is diagonal.

Theorem 3.2.1   Define E^m as in (3.1.10) via (3.1.5) and (3.1.7). Then

$$E^m = \text{diag}(e_{00}^m, \cdots, e_{K(m)-1,K(m)-1}^m). \qquad (3.2.1)$$

Proof.   From (2.1.20), the Fourier expansion of function $A_k^{n,m}$ can be written into

$$A_k^{n,m}(x) = \sum_{l \in \mathbb{Z}} q_{k+lK(m)} e^{i(k+lK(m))x}, \qquad (3.2.2)$$

and

$$\theta(x) = \sum_{k \in \mathbf{Z}} p_k e^{ikx} \quad \text{(see (3.1.6))} \tag{3.2.3}$$

$$e_{\mu,\nu}^m = \frac{1}{2\pi} \int_0^{2\pi} \int_0^{2\pi} \theta(y-t) A_\nu^{n,m}(t) \overline{A}_\mu^{n,m}(y) dt dy$$

$$= \frac{1}{2\pi} \sum_{k \in \mathbf{Z}} p_k \sum_{\lambda_1, \lambda_2 \in \mathbf{Z}} q_{\lambda_1} \overline{q}_{\lambda_2}$$

$$\times \int_0^{2\pi} \int_0^{2\pi} e^{i(k-\mu-\lambda_2 K(m))y + i(\nu - k + \lambda_1 K(m))t} dy dt$$

$$= \frac{1}{2\pi} \sum_{k \in \mathbf{Z}} p_k \sum_{\lambda_1 \lambda_2 \in \mathbf{Z}} q_{\lambda_1} \overline{q}_{\lambda_2} (2\pi)^2 \delta_{k,\mu+\lambda_2 K(m)} \delta_{k,\nu+\lambda_1 K(m)}$$

$$= \left(2\pi \cdot \sum_{\lambda \in \mathbf{Z}} p_{\mu+\lambda K(m)} |q_\lambda|^2 \right) \delta_{\nu,\mu} = e_{\mu,\mu}^m \delta_{\nu,\mu}.$$

□

Although the proof of this theorem is not very long, yet it is useful in the application.

Remove the term $E_{dd}^{m-1} d_{m-1}^m$ to the left side of (3.1.13) and then multiply both sides by $\gamma^{m-1} := (I - E_{dd}^{m-1})^{-1}$,

$$d_{m-1}^m = \gamma^{m-1}(E_{ds}^{m-1} + F_{ds}^{m-1})S_{m-1}^m + \gamma^{m-1} F_{dd}^{m-1} d_{m-1}^m + \gamma^{m-1} g_d^{m-1} \tag{3.2.4}$$

where we assume that γ^{m-1} exists for all large m.

Substituting (3.2.4) into (3.1.12), we obtain

$$S_{m-1}^m = (\tilde{E}^{m-1} + \tilde{F}^{m-1})S_{m-1}^m + \rho^{m-1} + \tilde{g}^{m-1} \tag{3.2.5}$$

where

$$\tilde{E}^{m-1} = E_{ss}^{m-1} + E_{sd}^{m-1} \gamma^{m-1} E_{ds}^{m-1}, \qquad \tilde{F}^{m-1} = F_{ss}^{m-1} \tag{3.2.6}$$

$$\rho^{m-1} = E_{sd}^{m-1} \gamma^{m-1} F_{ds}^{m-1} S_{m-1}^m + F_{sd}^{m-1} d_{m-1}^m$$

$$+ E_{sd}^{m-1} \gamma^{m-1} F_{dd}^{m-1} d_{m-1}^m \tag{3.2.7}$$

$$\tilde{g}^{m-1} = g_s^{m-1} + E_{sd}^{m-1} \gamma^{m-1} g_d^{m-1}. \tag{3.2.8}$$

Later we shall show that $E_\&^{m-1}$ has the order h_m and $F_\&^{m-1}$ has order h_m^r, where $r(r \geq 2)$ is the smooth order of $b(x,y)$. Therefore all the terms in ρ^{m-1} are of order h_m^r; neglecting the term ρ^{m-1} the new solution is denoted by $\tilde{S}^{m-1}$. Then (3.2.5) has the following form

$$\tilde{S}^{m-1} = (\tilde{E}^{m-1} + \tilde{F}^{m-1})\tilde{S}^{m-1} + \tilde{g}^{m-1}. \qquad (3.2.9)$$

Comparing (3.2.9) with (3.1.11), we find that the number of unknowns in (3.2.9) is half of that in (3.1.11), since $\tilde{S}^{m-1}$ is a $K(m-1) \times 1$ column and S^m is of size $K(m) \times 1, K(m) = 2K(m-1)$.

We shall prove that $\tilde{S}^{m-1}$ and S_{m-1}^m are slightly different in section 6 (see Lemma 3.6.5).

If we have $\tilde{S}^{m-1}$ then we need to have d_{m-1}^m to reconstruct S^m. The value of d_{m-1}^m can be approximated by

$$\tilde{d}^{m-1} = \gamma^{m-1} E_{ds}^{m-1} \tilde{S}^{m-1} + \gamma^{m-1} g_d^{m-1}. \qquad (3.2.10)$$

(3.2.10) is obtained from (3.2.4) by throwing away the two small terms F_{dd}^{m-1} and F_{ds}^{m-1}, and using $\tilde{S}^{m-1}$ instead of S_{m-1}^m.

Multiply W_{m-1} to both sides of (3.2.9) and split into a pair of equations to obtain

$$\tilde{S}_{m-2}^{m-1} = (\tilde{E}_{ss}^{m-2} + \tilde{F}_{ss}^{m-2})\tilde{S}_{m-2}^{m-1} + (\tilde{E}_{sd}^{m-1} + \tilde{F}_{sd}^{m-2})\tilde{d}_{m-2}^{m-1} + \tilde{g}_s^{m-2} \quad (3.2.11)$$

$$\tilde{d}_{m-2}^{m-1} = (\tilde{E}_{ds}^{m-2} + \tilde{F}_{ds}^{m-2})\tilde{S}_{m-2}^{m-1} + (\tilde{E}_{dd}^{m-2} + \tilde{F}_{dd}^{m-2})\tilde{d}_{m-2}^{m-1} + \tilde{g}_d^{m-2} \quad (3.2.12)$$

where we write $\tilde{S}_{m-2}^{m-1}$ instead of $L_{m-1}\tilde{S}^{m-1}$, etc. Move the term $\tilde{E}_{dd}^{m-2}\tilde{d}_{m-2}^{m-1}$ to the left of (3.2.12), multiply both sides by $\tilde{\gamma}^{m-1}$, substitute the value of $\tilde{d}_{m-1}^{m-1}$ into (3.2.11) to obtain

$$\tilde{S}_{m-2}^{m-1} = (\tilde{E}^{m-2} + \tilde{F}^{m-2})\tilde{S}_{m-2}^{m-1} + \tilde{g}^{m-2} + \tilde{\rho}^{m-2} \qquad (3.2.13)$$

where

$$\tilde{E}^{m-2} = \tilde{E}_{ss}^{m-2} + \tilde{E}_{sd}^{m-1}\tilde{\gamma}^{m-2}\tilde{E}_{ds}^{m-2}, \quad \tilde{F}^{m-2} = \tilde{F}_{ss}^{m-2},$$
$$\tilde{g}^{m-2} = \tilde{g}_s^{m-2} + \tilde{E}_{sd}^{m-2}\tilde{\gamma}^{m-2}\,\tilde{g}_d^{m-2},$$
$$\tilde{\rho}^{m-2} = \tilde{E}_{sd}^{m-2}\tilde{\gamma}^{m-2}\tilde{F}_{ds}^{m-2}\tilde{S}_{m-2}^{m-1} + \tilde{F}_{sd}^{m-2}\tilde{d}_{m-2}^{m-1} + \tilde{E}_{sd}^{m-2}\tilde{\gamma}^{m-2}\tilde{F}_{dd}^{m-2}\tilde{d}_{m-2}^{m-1},$$

and

$$\tilde{\gamma}^{m-2} = (I - \tilde{E}_{dd}^{m-2})^{-1},$$

I is an identity matrix of size $K(m-2) \times K(m-2)$. If we omit the terms of higher order, write $\tilde{S}^{m-2}$, $\tilde{d}^{m-2}$ and $\tilde{g}^{m-2}$ instead of $\tilde{S}_{m-2}^{m-1}$, $\tilde{d}_{m-2}^{m-1}$ and $\tilde{g}_s^{m-2}$ respectively, then (3.2.11) and (3.2.12) can be written into the following equations

$$\tilde{S}^{m-2} = (\tilde{E}^{m-2} + \tilde{F}^{m-2})\tilde{S}^{m-2} + \tilde{g}^{m-2} \qquad (3.2.14)$$

$$\tilde{d}^{m-2} = \tilde{\gamma}^{m-2}\tilde{E}_{ds}^{m-2}\tilde{S}^{m-2} + \tilde{\gamma}^{m-2}\tilde{g}_d^{m-2} \qquad (3.2.15)$$

In general, if we have the following equation

$$\tilde{S}_k^{k+1} = (\tilde{E}^k + \tilde{F}^k)\tilde{S}_k^{k+1} + \tilde{g}^k + \rho^k \qquad (3.2.16)$$

where

$$\tilde{S}_k^{k+1} = L_{k+1}\tilde{S}^{k+1}, \quad \tilde{d}_k^{k+1} = H_{k+1}\tilde{S}^{k+1},$$

$$\tilde{E}^k = \tilde{E}_{ss}^k + \tilde{E}_{sd}^k\tilde{\gamma}^k\tilde{E}_{ds}^k, \quad \tilde{F}^k = \tilde{F}_{ss}^k,$$

$$\tilde{E}_{ss}^k = L_{k+1}\tilde{E}^{k+1}L_{k+1}^T, \quad \tilde{E}_{sd}^k = L_{k+1}\tilde{E}^{k+1}H_{k+1}^T,$$

$$\tilde{E}_{ds}^k = H_{k+1}\tilde{E}^{k+1}L_{k+1}^T, \quad \tilde{E}_{dd}^k = H_{k+1}\tilde{E}^{k+1}H_{k+1}^T, \qquad (3.2.17)$$

$$\tilde{F}_{ss}^k = L_{k+1}\tilde{F}^{k+1}L_{k+1}^T, \quad \tilde{F}_{sd}^k = L_{k+1}\tilde{F}^{k+1}H_{k+1}^T,$$

$$\tilde{F}_{ds}^k = H_{k+1}\tilde{F}^{k+1}L_{k+1}^T, \quad \tilde{F}_{dd}^k = H_{k+1}\tilde{F}^{k+1}H_{k+1}^T,$$

$$\tilde{\gamma}^k = (I - \tilde{E}_{dd}^k)^{-1}, \quad (I \text{ is an } K(k) \times K(k) \text{ identity matrix}),$$
$$\qquad (3.2.18)$$

$$\tilde{g}^k = \tilde{g}_s^k + \tilde{E}_{sd}^k\tilde{\gamma}^k\tilde{g}_d^k, \quad \tilde{g}_s^k = L_{k+1}\tilde{g}^{k+1}, \quad \tilde{g}_d^k = H_{k+1}\tilde{g}^{k+1}$$

and

$$\rho^k = \tilde{E}_{sd}^k\tilde{\gamma}^k\tilde{F}_{ds}^k\tilde{S}_k^{k+1} + \tilde{E}_{sd}^k\tilde{\gamma}^k\tilde{F}_{dd}^k\tilde{d}_k^{k+1} + \tilde{F}_{sd}^k\tilde{d}_k^{k+1} \qquad (3.2.19)$$

since the norm of ρ^k is small, we omit ρ^k in (3.2.16), replace $\tilde{S}_k^{k+1}$ by $\tilde{S}^k$, then from (3.2.16) we have

$$\tilde{S}^k = (\tilde{E}^k + \tilde{F}^k)\tilde{S}^k + \tilde{g}^k \qquad (3.2.20)$$

From (3.2.10), we also have

$$\tilde{d}^k = \tilde{\gamma}^k \tilde{E}_{ds}^k \tilde{S}^k + \tilde{\gamma}^k \tilde{g}_d^k. \tag{3.2.21}$$

We now set $\tilde{E}^m = E^m$, $\tilde{F}^m = F^m$, $\tilde{g}^m = g^m$ and $\tilde{S}^m = S^m$. Then (3.2.20) is valid for k from $k = m$ up to $k = m_1$, $m_1 < m$, m_1 some positive integer.

§3.3   Algorithm

Let $k = m_1$. Then we can solve $\tilde{S}^{m_1}$ from

$$\tilde{S}^{m_1} = (\tilde{E}^{m_1} + \tilde{F}^{m_1})\tilde{S}^{m_1} + \tilde{g}^{m_1} \tag{3.3.1}$$

by the Gaussian elimination method.

Define

$$\overline{d}^k = \tilde{\gamma}^k \tilde{E}_{ds}^k \overline{S}^k + \tilde{\gamma}^k \tilde{g}_d^k \tag{3.3.2}$$

$$\overline{S}^{k+1} = L_{k+1}^T \overline{S}^k + H_{k+1}^T \overline{d}^k \tag{3.3.3}$$

Let $k = m_1$ in (3.3.2). We obtain $\overline{d}^{m_1}$ and substitute it into (3.3.3), then we have $\overline{S}^{m_1+1}$, etc.

The whole procedure is as follows:

Step 1. Precompute E^m and F^m (see (3.1.10)) for some sufficiently large number m. Precompute $\tilde{E}^k$, $\tilde{F}^k$, $\tilde{E}_l^k$ and $\tilde{F}_l^l$, for $m_1 \le k < m$ by using the formulas (3.2.17), where l stands for ss, sd, ds or dd.

Step 2. Compute $\tilde{g}^k, \tilde{g}_s^k$ and $\tilde{g}_d^k$ for $m_1 \le k < m$ and g^m by using (3.2.18) and (3.1.10). $\hfill (3.3.4)$

Step 3. Solve the linear system (3.2.20) for $k = m_1$.

Step 4. Compute $\overline{d}^k$ from (3.3.2), $\overline{S}^{m_1} := \tilde{S}^{m_1}$.

Step 5. Reconstruct $(\overline{S}^k, \overline{d}^k) \to \overline{S}^{k+1}$ by using (3.3.3). Let $k + 1 \to k$.

Step 6. Go to Step 4 until $k = m$.

When we complete all these steps, we can compute the approximate solution of (3.1.2) by using (3.1.3) with $\overline{S}^m$ instead of $\tilde{S}^m$, we thus obtain an approximate solution of u_m :

$$\bar{u}_m(x) = \sum_{k=0}^{K(m)-1} \overline{S}_k^m A_k^{n,m}(x) \tag{3.3.5}$$

where we denote $\overline{S}^m = (\overline{S}_k^m)$.

§3.4   Complexity

We consider the complexity of the computation given above. As usual (see [BCR], [CMX] and [DKPS]), we assume that the stiffness matrix is given that means the computations in the Step 1 have been completed. When we solve a large number of integral equations with the same integral kernel and different $g(x)$, we need compute the quantities in the first step (Step 1) for one time and reuse these numbers at every different equations.

In Step 2 we have to compute $\tilde{g}^k, \tilde{g}_s^k$ and $\tilde{g}_d^k$ for $k : m_1 \le k < m$. g^m is precomputed, there are $2K(m)$ multiplications to get $\tilde{g}_s^{m-1}$ from $\tilde{g}^m$ by using the formula $\tilde{g}_s^{m-1} = L_m \tilde{g}^m (\tilde{g}^m = g^m)$, and $\tilde{g}_d^{m-1} = H_m \tilde{g}^m$. Since $\tilde{g}^{m-1} = \tilde{g}_s^{m-1} + \tilde{E}_{sd}^{m-1} \tilde{\gamma}^{m-1} \tilde{g}_d^{m-1}$, there are $2K(m-1)$ multiplications to get $\tilde{g}^{m-1}$. Each of the calculation for $\tilde{g}_s^{m-2}$ and $\tilde{g}_d^{m-2}$ from $\tilde{g}^{m-1}$ contains $2K(m-1)$ multiplications, etc.. We can graph as follows:

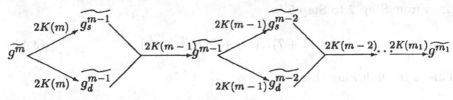

the number of the total operations is less than $11K(m)$. In fact, we have

$$4K(m) \quad + 6K(m-1) + \cdots + 6K(m_1+1) + 2K(m_1)$$

$$\le 4K(m) + 12K(m-1) + 2K(m_1) \tag{3.4.1}$$
$$\le 10K(m) + 2K(m_1)$$
$$< 11K(m).$$

In Step 3 it takes $O(K^3(m_1))$ operations by Gaussian elimination to solve the equation $\tilde{S}^{m_1} = (\tilde{E}^{m_1} + \tilde{F}^{m_1})\tilde{S}^{m_1} + \tilde{g}^{m_1}$. Since

$$K^3(m_1) = 2^{3m_1}K^3 < 2^{m \cdot \frac{3m_1}{m}} \cdot K^{\frac{3m_1}{m}} = (K(m))^{\frac{3m_1}{m}},$$

$O(K^3(m_1))$ is less than $O(K(m)^{\frac{3m_1}{m}})$.

In Step 4, to compute $\overline{d}^{m_1}$ from the formula

$$\overline{d}^{m_1} = \tilde{\gamma}^{m_1}\tilde{E}_{ds}^{m_1}\overline{S}^{m_1} + \tilde{\gamma}^{m_1}\tilde{g}_d^{m_1}$$

it takes $3K(m_1)$ operations.

In Step 5, by using

$$\overline{S}^{k+1} = L_{k+1}^T\overline{S}^k + H_{k+1}^T\overline{d}^k,$$

to calculate $\overline{S}^{k+1}$, from $(\overline{S}^k, \overline{d}^k)$ it takes $4K(k)$ operations. Let $k+1 \to k$. Then go back to Step 4. The amount of operations for k from m_1 to $m-1$ is

$$\sum_{k=m_1}^{m-1}(3K(k) + 4K(k)) = 7\sum_{k=m_1}^{m-1}K(k)$$
$$\leq 14K(m-1) \tag{3.4.2}$$
$$= 7K(m).$$

(3.4.1) and (3.4.2) together, we conclude that the total number of operations from Step 2 to Step 5 is

$$(11+7)K(m) + O(K(m)^{\frac{3m_1}{m}}). \tag{3.4.3}$$

This is the following theorem

Theorem 3.4.1   Assume that the stiffness matrix is given and $m_1 = \left[\frac{m+2}{3}\right]$, then the complexity of the algorithm is $O(K(m))$.

Proof.   Since $3m_1 = 3\left[\frac{m+2}{3}\right] \leq m+2$, from (3.4.3) we obtain the conclusion of Theorem 3.4.1.   □

§3.5   The Convergence of the Approximate Solution

Before we quote some lemmas we recall the following definition of compact operators.

Definition 3.5.1   A subset U of a normed space X is called compact if every open covering of U contains a finite subcovering, that is, if for every family V_j, $j \in J$, of open sets with the property

$$U \subset \bigcup_{j \in J} V_j$$

there exists a finite subfamily $V_{j(k)}$, $j(k) \in J, k = 1, \cdots, n$, with

$$U \subset \bigcup_{k=1}^{n} V_j(k).$$

Definition 3.5.2   A linear operator $A : X \to Y$ from a normed space X into a normed space Y is called compact if it maps each bounded set in X into a relatively compact set in Y.

Lemma 3.5.1   Let X be a normed space, $A : X \mapsto X$ a compact operator, and let $I - A$ be injective (i.e.,the null space of $I - A : N(I - A) = \{0\}$). Then the inverse operator $(I - A)^{-1} : X \mapsto X$ exists and is bounded.

Lemma 3.5.2   Assume that A has a bounded inverse $A^{-1} : Y \mapsto X$. Then for sufficiently large n, more precisely, for all n with

$$\|A^{-1}(A_n - A)\| < 1$$

the inverse operators $A_n^{-1} : Y \mapsto X$ exist and are uniformly bounded by

$$\|A_n^{-1}\| \leq \frac{\|A^{-1}\|}{1 - \|A^{-1}(A_n - A)\|}.$$

(refer to [Kr, Chapter 3 and Chapter 10]).

Lemma 3.5.3   Let

$$T_1 u(x) = \int_0^{2\pi} a(x-y)\ln\left|2\sin\frac{x-y}{2}\right| u(y)dy,$$

where $u \in \overset{\circ}{C}{}^s ([0,2\pi])$. Then

$$\|T_1 u - P_m T_1 u\|_{\overset{\circ}{L}_2} \le Ch_m^S \qquad (3.5.1)$$

where C is a constant, and the $\overset{\circ}{L}_2$ norm is on the space $\overset{\circ}{L}_2 ([0,2\pi])$.

Proof. $T_1 u \in \overset{\circ}{C}{}^s ([0,2\pi])$ since $u \in \overset{\circ}{C}{}^s ([0,2\pi])$, then (3.5.1) follows from [Sc. p.307–308].   □

Theorem 3.5.1   Assuming that u_m is the solution of (3.1.2), u the solution of (3.1.1). We also assume $I - T$ has bounded inverse, $b \in \overset{\circ}{C}{}^r$ $([0,2\pi] \times [0,2\pi])$, $a, g \in \overset{\circ}{C}{}^s ([0,2\pi])$, and $u \in \overset{\circ}{C}{}^s ([0,2\pi])$, where $r, s \le n$, and $s \le r$. Then

$$\|u - u_m\| \le Ch_m^s. \qquad (3.5.2)$$

where C is a constant independent of m.

Proof.   Because u and u_m satisfies (3.1.1) and (3.1.2) respectively,

$$u - u_m = Tu - P_m T u_m + g - P_m g$$
$$\qquad\qquad\qquad\qquad (3.5.3)$$
$$= Tu - P_m Tu + P_m Tu - P_m T u_m + g - P_m g.$$

From Lemma 3.5.2, $I - P_m T$ is invertible for large m, in this case, their inverses have same bound. By using Lemma 3.5.3 and (2.1.55),

$$\|u - u_m\|_{\overset{\circ}{L}_2([0,2\pi])} = \|(I - P_m T)^{-1}\| \, \|(Tu - P_m Tu) + (g - P_m g)\| \le Ch_m^s$$

where we use the smoothness of $b(x,y)$ in applying Theorem 2.1.7.

Theorem 3.5.1 follows immediately.   □

Remark   If $I - T$ has null space $\{0\}$, by Lemma 3.5.1, $(I - T)^{-1}$ exists since T is a compact operator, therefore the condition in Theorem 3.5.1 will be satisfied. For those integral equations induced from Laplace equations or conformal mapping problems, $I - T$ has trivial null space (see [KS1]), hence we can use the estimate (3.5.2).

In §3.3, we got an approximate solution of u_m, i.e., $\bar{u}_m$ (see (3.3.5)). In order to estimate the error, we shall study some error analysis.

§3.6   Error Analysis

In this section, we shall analyze the computational errors in the algorithm. We shall use the notations defined in previous sections.

Recall S^m is the solution of the following equation (see (3.1.11)).

$$S^m = (E^m + F^m)S^m + g^m. \tag{3.6.1}$$

Let $\overline{S}^m$ be the approximate solution of S^m, which is defined in (3.3.3) for $k = m - 1$. We need to estimate the error $\|S^m - \overline{S}^m\|_{l^2}$.

Since $L_m^T L_m + H_m^T H = I_m$, from (3.1.15) and (3.3.3), we have

$$S^m = L_m^T S_{m-1}^m + H_m^T d_{m-1}^m, \tag{3.6.2}$$

$$\overline{S}^m = L_m^T \overline{S}^{m-1} + H_m^T \overline{d}^{m-1}. \tag{3.6.3}$$

Since $W_m = \begin{pmatrix} L_m \\ H_m \end{pmatrix}$ is unitary,

$$\|S^m - \overline{S}^m\|_{l^2}^2 = \|L_m^T(S_{m-1}^m - \overline{S}^{m-1}) + H_m^T(d_{m-1}^m - \overline{d}^{m-1})\|_{l^2}^2 \tag{3.6.4}$$

$$= \|S_{m-1}^m - \overline{S}^{m-1}\|_{l^2}^2 + \|d_{m-1}^m - \overline{d}^{m-1}\|_{l^2}^2$$

Since the difference $S_{m-1}^m - \tilde{S}^{m-1}$ is ρ^{m-1} (see (3.2.5) and (3.2.9)), ρ^{m-1} is a small quantity, therefore we may estimate the quantity $\|\tilde{S}^{m-1} - \overline{S}^{m-1}\|_{l^2}$ instead of $\|S_{m-1}^m - \overline{S}^{m-1}\|_{l^2}$. Before doing this we need to establish some lemmas.

Lemma 3.6.1 Under the conditions in Theorem 3.5.1, we have

$$\|E_{\&}^{k}\| \leq M_e h_k \tag{3.6.5}$$

and

$$\|F_{\&}^{k}\| \leq M_f h_k^r \tag{3.6.6}$$

where $\& = sd, ds$ or dd, k is any positive integer and M_e, M_f are constants independent of k.

Proof. We prove (3.6.6) first. We shall establish the following inequality

$$\|F_{\&}^{k}\| \leq \|b_{k+1} - b_k\|_{\overset{\circ}{L_2(I \times I)}} \quad \text{for} \quad \& = sd, ds, dd \tag{3.6.7}$$

where $b_k(x, y)$ is the projection of $b(x, y)$ onto $V_k \times V_k$, $b(x, y)$ is the function in the integral equation $(3.0)_1$, and $I = [0, 2\pi]$.

Let $F^{k+1} = (f_{lt}^{k+1})$ be a matrix of size $K(k+1) \times K(k+1)$, where

$$f_{l,t}^{k+1} = \frac{1}{2\pi} \int_0^{2\pi} \int_0^{2\pi} b(x, y) \overline{A_l^{n,k+1}}(x) A_t^{n,k+1}(y) dx dy \quad \text{(refer to } (3.1.9)_2) \tag{3.6.8}$$

$0 \leq l, t \leq K(k)$.

$$F_{sd}^{k} = L_{k+1} F^{k+1} H_{l+1}^T = (F_{sd}^{k}(l, t))$$

is a $K(k) \times K(k)$ matrix, where

$$F_{sd}^{k}(l, t) = a_l^{n,k+1} f_{l,t}^{k+1} b_t^{n,k+1} + a_l^{n,k+1} f_{l,t+K(k)}^{k+1} (-a_t^{n,k+1})$$

$$+ b_l^{n,k+1} f_{l+K(k),t}^{k+1} b_t^{n,k+1} + b_l^{n,k+1} f_{l+K(k),t+K(k)}^{k+1} (-a_t^{n,k+1})$$

$$= \frac{1}{2\pi} \int_0^{2\pi} \int_0^{2\pi} b(x,y) \{ a_l^{n,k+1} \overline{A}_l^{n,k+1}(x) A_t^{n,k+}(y) b_t^{n,k+1}$$

$$+ a_l^{n,k+1} \overline{A}_l^{n,k+1}(x) A_{t+K(k)}^{n,k+1}(y)(-a_t^{n,k+1})$$

$$+ b_l^{n,k+1} \overline{A}_{l+K(k)}^{n,k+1}(x) A_t^{n,k+1}(y) b_t^{n,k+1}$$

$$+ b_l^{n,k+1} \overline{A}_{l+K(k)}^{n,k+1}(x) A_{t+K(k)}^{n,k+1}(y)(-a_t^{n,k+1}) \} dx dy$$

$$= \frac{1}{2\pi} \int_0^{2\pi} \int_0^{2\pi} b(x,y) \{ a_l^{n,k+1} \overline{A}_l^{n,k+1}(x) + b_l^{n,k+1} \overline{A}_{l+K(k)}^{n,k+1}(x) \}$$

$$\times \{ A_t^{n,k+1}(y) b_t^{n,k+1} + A_{t+K(k)}^{n,k+1}(y)(-a_l^{n,k+1}) \} dx dy$$

$$= -\frac{1}{2\pi} \int_0^{2\pi} \int_0^{2\pi} b(x,y) \overline{A}_l^{n,k}(x) D_t^{n,k}(y) dx dy \quad \text{(refer to Chapter II)}$$

$$= -2\pi \langle b(x,y), A_l^{n,k}(x) \overline{D}_t^{n,k}(y) \rangle. \tag{3.6.9}$$

We can obtain analogous formulas for $F_{\&}^k(l,t)$, where $\& = ds, dd$ and ss.

Since

$$V_{k+1} \times V_{k+1} = V_k \times V_k + V_k \times W_k + W_k \times V_k + W_k \times W_k,$$

we have

$$b_{k+1}(x,y) = \sum_{l=0}^{K(k)-1} \sum_{t=0}^{K(k)-1} \langle b(x,y), A_l^{n,k}(x) \overline{A}_t^{n,k}(y) \rangle A_l^{n,k}(x) \overline{A}_t^{n,k}(y)$$

$$+ \sum_{l=0}^{K(k)-1} \sum_{t=0}^{K(k)-1} \langle b(x,y), A_l^{n,k}(x) \overline{D}_t^{n,k}(y) \rangle A_l^{n,k}(x) \overline{D}_t^{n,k}(y)$$

$$+ \sum_{l=0}^{K(k)-1} \sum_{t=0}^{K(k)-1} \langle b(x,y), D_l^{n,k}(x) \overline{A}_t^{n,k}(y) \rangle D_l^{n,k}(x) \overline{A}_t^{n,k}(y)$$

$$+ \sum_{l=0}^{K(k)-1} \sum_{t=0}^{K(k)-1} \langle b(x,y), D_l^{n,k}(x) \overline{D}_t^{n,k}(y) \rangle D_l^{n,k}(x) \overline{D}_t^{n,k}(y)$$

$$= \frac{1}{2\pi} \sum_{l=0}^{K(k)-1} \sum_{t=0}^{K(k)-1} \{F_{ss}^k(l,t) A_l^{n,k}(x) \overline{A}_t^{n,k}(y)$$

$$+ F_{sd}^k(l,t) A_l^{n,k}(x) \overline{D}_t^{n,k}(y)$$

$$- F_{ds}^k(l,t) D_l^{n,k}(x) \overline{A}_t^{n,k}(y)$$

$$+ F_{dd}^k(l,t) D_l^{n,k}(x) \overline{D}_l^{n,k}(y)\}. \tag{3.6.10}$$

The norm of the matrix F_{sd}^k is denoted by $\|F_{sd}^k\|$. Then

$$\|F_{sd}^k\|^2 \leq \sum_l \sum_t (F_{sd}^k(l,t))^2$$

$$\leq \sum_l \sum_t \{(F_{sd}^k(l,t))^2 + (F_{ds}^k(l,t))^2 + (F_{dd}^k(l,t))^2\}$$

$$= \frac{1}{4\pi^2} \|b_{k+1} - b_k\|_{\overset{\circ}{L_2(I \times I)}}^2$$

$$\leq \frac{1}{4\pi^2} \{\|b_{k+1} - b_k\|_{\overset{\circ}{L_2(I \times I)}}^2 + \|b - b_{k+1}\|_{\overset{\circ}{L_2(I \times I)}}^2\} \tag{3.6.11}$$

$$= \frac{1}{4\pi^2} \|b - b_{k+1} + b_{k+1} - b_k\|_{\overset{\circ}{L_2(I \times I)}}^2$$

$$= \frac{1}{4\pi^2} \|b - b_{k+1}\|_{\overset{\circ}{L_2(I \times I)}}^2.$$

Since $b_{k+1} - b_k \in V_{k+1} \times V_{k+1}$, $b - b_{k+1} \in (V_{k+1} \times V_{k+1})^\perp$.

Inequality (3.6.6) for $\& = sd$ follows from $b \in \overset{\circ}{C}{}^r (I \times I)$, (3.6.11) and (2.1.55). As for $\& = ds, dd$, the similar method can be applied.

For the proof of (3.6.5), we only consider the case $\& = sd$ since the others are similar.

Since E_{sd}^k is diagonal we need to calculate $E_{sd}^k(j,j)$ only.

As we derived (3.6.9), we can easily obtain

$$E_{sd}^k(j,j) = \frac{a_j^{n,k+1} b_j^{n,k+1}}{2\pi} \int_0^{2\pi} \int_0^{2\pi} \theta(x-y)$$

$$\times \{\overline{A_j^{n,k+1}}(x) A_j^{n,k+1}(y) - \overline{A_{j+K(k)}^{n,k+1}}(x) A_{j+K(k)}^{n,k+1}(y)\} dx dy \tag{3.6.12}$$

$$\theta(x - y) = a(x - y) \ln \left| 2 \sin \frac{x - y}{2} \right|$$

$$a(x - y) = a_0 + a_1(x - y) \sin^2 \frac{x - y}{2} \qquad (3.6.13)$$

$$a_1(t) \in \overset{\circ}{C^1}$$

If $\theta(x)$ has a Fourier expansion $\theta(x) = \sum_k C_k e^{-ikx}$, we prove that

$$|C_k| \le \frac{A}{|k|}, \quad k \ne 0, \quad \forall k \in \mathbf{Z}, \qquad (3.6.14)$$

where A is a constant.

$$
\begin{aligned}
C_k &= \frac{1}{2\pi} \int_0^{2\pi} \theta(t) e^{ikt} dx \\
&= \frac{1}{2\pi} \int_0^{2\pi} \left(a_0 + a_1(t) \sin^2 \frac{t}{2} \right) \ln \left| 2 \sin \frac{t}{2} \right| e^{ikt} dt \\
&= \frac{1}{2\pi} \int_0^{2\pi} a_0 \ln \left| 2 \sin \frac{t}{2} \right| e^{ikt} dt + \frac{1}{2\pi} \int_0^{2\pi} a_1(t) \left(\sin^2 \frac{t}{2} \right) \ln \left| 2 \sin \frac{t}{2} \right| e^{ikt} dt \\
&= -\frac{a_0}{2} \cdot \frac{1}{|k|} + I_1. \qquad (3.6.15)
\end{aligned}
$$

We hence only need to prove

$$|I_1| \le \frac{A_0}{|k|}, \qquad (3.6.16)$$

where A_0 is a constant. Set $\tilde{a}(t) = a_1(t) \left(\sin^2 \frac{t}{2} \right) \ln \left| 2 \sin \frac{t}{2} \right|$, then $\tilde{a}(t) \in \overset{\circ}{C^1}$, we have

$$
\begin{aligned}
|I_1| &= \left| \frac{1}{2\pi} \int_0^{2\pi} \tilde{a}(t) e^{ikt} dt \right| \\
&= \left| \frac{-1}{2\pi} \int_0^{2\pi} \frac{1}{ik} \tilde{a}'(t) e^{ikt} dt \right| \\
&\le \frac{1}{|k|} \cdot \frac{1}{2\pi} \int_0^{2\pi} |\tilde{a}'(t)| |e^{ikt}| dt \\
&\le \frac{1}{|k|} \sup_{t \in [0, 2\pi]} |\tilde{a}'(t)|,
\end{aligned}
\qquad (3.6.17)
$$

hence we have (3.6.16), where $A_0 = \sup_{t \in [0, 2\pi]} |\tilde{a}'(t)|$. (3.6.14) follows from (3.6.15) and (3.6.16). By the Fourier expansion of $A_j^{n,k}(x)$ (see (2.1.20)), and using (3.6.14) we have

$$\omega_1 := \left| \int_0^{2\pi} \int_0^{2\pi} \theta(x - y) \overline{A}_j^{n,k+1}(x) A_j^{n,k+1}(y) dx dy \right|$$

$$= 4\pi^2 (C_j^{n,k+1})^2 (K(k+1))^{2n+2}$$

$$\times \left| \sum_{\lambda \in \mathbb{Z}} \left(\frac{\sin \frac{j\pi}{K(k+1)}}{(j + \lambda K(k+1))\pi} \right)^{2n+2} C_{-j-\lambda K(k+1)} \right|$$

$$\leq \hat{C}_1 \left\{ \frac{A}{j} + (K(k+1))^{2n+2} \right.$$

$$\left. \times \sum_{\substack{\lambda \in \mathbb{Z} \\ \lambda \neq 0}} \frac{1}{(j + \lambda K(k+1))^{2n+2}} \cdot \frac{1}{|j + \lambda K(k+1)|} \right\}$$

$$= \hat{M}_1(k+1, j)$$

$$(3.6.18)$$

where $\hat{C}_1$ is a constant.

Now we estimate $\hat{M}_1(k+1, j)$.

Assume $0 < j < K(k)$. Let

$$\frac{\hat{M}_1(k+1, j)}{\hat{C}_1} = M_1 + M_2 + M_3 + M_4, \qquad (3.6.19)$$

where

$$M_1 = (K(k+1))^{2n+2} \sum_{\lambda \leq -2, \lambda \in \mathbb{Z}} \frac{1}{(j + \lambda K(k+1))^{2n+2}} \cdot \frac{1}{|j + \lambda K(k+1)|}$$

$$= \frac{1}{K(k+1)} \sum_{\lambda \geq 2, \lambda \in \mathbb{Z}} \frac{1}{(\lambda - \frac{j}{K(k+1)})^{2n+3}}$$

$$< \frac{1}{K(k+1)} \sum_{\lambda \geq 2, \lambda \in \mathbf{Z}} \frac{1}{(\lambda - 1)^{2n+3}}$$

$$\leq \frac{A_1}{K(k+1)} \tag{3.6.20}$$

$$< \frac{A_1}{j},$$

A_1 a constant.

$$M_2 = (K(k+1))^{2n+2} \frac{1}{j(K(k+1) - j)^{2n+3}}$$

$$= \frac{1}{K(k+1) - j} \cdot \left(\frac{K(k+1)}{K(k+1) - j} \right)^{2n+2}$$

$$< \frac{1}{j} \cdot 2^{2n+2} \quad \text{(since } 0 < j < K(k))$$

$$\tag{3.6.21}$$

$$M_3 = (K(k+1))^{2n+2} \sum_{\lambda > 0, \lambda \in \mathbf{Z}} \frac{1}{(j + \lambda K(k+1))^{2n+3}}$$

$$= \frac{1}{K(k+1)} \cdot \sum_{\lambda > 0, \lambda \in \mathbf{Z}} \frac{1}{(\lambda + \frac{j}{K(k+1)})^{2n+3}}$$

$$< \frac{1}{K(j+1)} \sum_{\lambda > 0, \lambda \in \mathbf{Z}} \frac{1}{\lambda^{2n+3}}$$

$$= \frac{A_2}{K(k+1)}$$

$$< \frac{A_2}{j}. \tag{3.6.22}$$

$$M_4 = \frac{A}{j}. \tag{3.6.23}$$

From (3.6.17)–(3.6.23), we have

$$\omega_1 = \frac{B_1}{j} \quad \text{for} \quad 0 < j < K(k) \tag{3.6.24}$$

where B_1 is a constant.

We now consider the second term in (3.6.12). Define

$$\omega_2 := \left| \int_0^{2\pi} \int_0^{2\pi} \theta(x - y) \overline{A}_{j+K(k)}^{n,k+1}(x) A_{j+K(k)}^{n,k+1}(y) dx dy \right|. \tag{3.6.25}$$

By using the analogous approach, we obtain the estimate

$$\omega_2 \leq \frac{B_2}{(K(k) - j)} \qquad \text{for} \quad 0 < j < K(k) \tag{3.6.26}$$

where B_2 is a constant.

Substituting $x = \dfrac{j\pi}{K(k+1)}$ into the inequality $|\cos x| \leq \left| \frac{\pi}{2} - x \right|$, we have

$$\left| \cos \frac{j\pi}{K(k+1)} \right| \leq \frac{(K(k) - j)\pi}{K(k+1)},$$

therefore

$$a_j^{n,k+1} \leq L_c \left(\frac{(K(k) - j)\pi}{K(k+1)} \right)^{n+1}$$

$$\tag{3.6.27}$$

$$b_j^{n,k+1} \leq L_c \left(\frac{j\pi}{K(k+1)} \right)^{n+1}, \quad L_c \text{ a constant.}$$

From (3.6.12), (3.6.24), (3.6.26) and (3.6.27), for $0 < j < K(k)$, we have

$$E_{s,d}^k(j,j) \leq \frac{a_j^{n,k+1} b_j^{n,k+1}}{2\pi} [\omega_1 + \omega_2]$$

$$\leq \frac{L_c^2}{2\pi} \frac{j(K(k) - j)\pi^2}{(K(k+1))^2} \left[\frac{B_1}{j} + \frac{B_2}{K(k) - j} \right]$$

$$= \frac{L_c^2}{2\pi(K(k+1))^2} [B_1 \pi^2 (K(k) - j) + B_2 \pi^2 j] \tag{3.6.28}$$

$$\leq \frac{B_3 K(k)}{(K(k+1))^2}$$

$$= \frac{B_3}{2} \frac{1}{K(k+1)} = \frac{B_4}{K(k+1)}$$

where B_3, B_4 are constants independent of k.

Since $b_0^{n,k+1} = 0$, we conclude from (3.6.12) that

$$E_{s,d}^k(0,0) = 0. \tag{3.6.29}$$

From (3.6.28), (3.6.29)

$$\|E_{sd}^k\| \leq \max_{0 \leq j < K(k)} \{E_{sd}^k(j,j)\} \leq \frac{C}{K(k+1)}$$

$$\leq \frac{C}{2\pi} \cdot \frac{2\pi}{2K(k)} = \frac{C}{4\pi} \cdot h_k$$

where C is a constant independent of k, (3.6.5) is proved for the case $\& = sd$, the others are similar.   □

In this section we declare that the conditions in Theorem 3.5.1 are satisfied.

The definitions of $\tilde{E}_\&^k$ and $\tilde{F}_\&^k$ are given in (3.2.17), and the definitions of $E_\&^k$ and $F_\&^k$ are given similarly in (3.1.16)–(3.1.18), the only difference is that m is replaced by k. Now we estimate the norm of $\tilde{E}_\&^k$ and $\tilde{F}_\&^k$, where $\&$ stands for sd, ds and dd, respectively.

Lemma 3.6.2   Let m_0 be an integer such that

$$M_e h_k < 0.25, \qquad \text{for} \quad k \geq m_0 \tag{3.6.30}$$

where M_e is the constant in (3.6.5).

If m is large enough, i.e., $m \geq 3m_0$, then

$$\|\tilde{E}_\&^k\| \leq 2M_e h_k \tag{3.6.31}$$

$$\|\tilde{F}_\&^k\| \leq 2M_f h_k^\tau \tag{3.6.32}$$

for

$$m > k \geq \max\left\{m_0, \left[\frac{m+2}{3}\right]\right\},$$

where $\& = sd, ds$ or dd.

Set $\tilde{\gamma}^k = (I - \tilde{E}_{dd}^k)^{-1}$, then we have

$$\|\tilde{\gamma}^k\| \leq 2 \tag{3.6.33}$$

Proof.   From (3.2.17) we have $\tilde{E}_{ss}^{m-1} = L_m \tilde{E}^m L_m^T$, by (3.1.16), $E_{ss}^{m-1} = L_m E^m L_m^T$, thus $\tilde{E}_{ss}^{m-1} = E_{ss}^{m-1}$ since $\tilde{E}^m := E^m$. Similarly we have $\tilde{E}_\&^{m-1} = E_\&^{m-1}$ and $\tilde{F}_\&^{m-1} = F_\&^{m-1}$. Thus (3.6.31) and (3.6.32) are true for $k = m - 1$.

Now suppose that (3.6.31), (3.6.32) are true for $k = k_0 (k_0 \leq m-1)$, we want to prove that they are also true for $k = k_0 - 1$.

From the sixth formula in (3.2.17)

$$\|\tilde{E}_{sd}^{k_0-1}\| = \|L_{k_0}\tilde{E}^{k_0}H_{k_0}^T\|$$

$$= \|L_{k_0}\tilde{E}_{ss}^{k_0}H_{k_0}^T + L_{k_0}\tilde{E}_{sd}^{k_0}\tilde{\gamma}^{k_0}\tilde{E}_{ds}^{k_0}H_{k_0}^T\|$$

$$\text{(by third formula in (3.2.17))}$$

$$\leq \|L_{k_0}\tilde{E}_{ss}^{k_0}H_{k_0}^T\| + \|\tilde{E}_{sd}^{k_0}\|\|\tilde{\gamma}^{k_0}\|\|\tilde{E}_{ds}^{k_0}\|$$

$$\text{(since } \|L_{k_0}\| \leq 1, \|H_{k_0}\| \leq 1)$$

$$\leq \|L_{k_0}\tilde{E}_{ss}^{k_0}H_{k_0}^T\| + 8(M_e h_{k_0})^2$$

$$(\|\tilde{\gamma}^{k_0}\| = \|(I - \tilde{E}_{dd}^{k_0})^{-1}\|$$

$$\leq \frac{1}{(1 - \|\tilde{E}_{dd}^{k_0}\|)} \leq 2, \quad \text{since by induction assumption}$$

$$\|\tilde{E}_{dd}^{k_0}\| \leq 2M_e h_{k_0} \leq \tfrac{1}{2})$$

$$\leq \|L_{k_0}L_{k_0+1}\tilde{E}^{k_0+1}L_{k_0+1}^T H_{k_0}^T\| + M_e h_{k_0}$$

$$\text{(from (3.34) and } M_e h_{k_0} = \tfrac{1}{2}M_e h_{k_0-1})$$

$$\leq \|L_{k_0}L_{k_0+1}\tilde{E}_{ss}^{k_0+1}L_{k_0+1}^T H_{k_0}^T\| + M_e h_{k_0+1} + M_e h_{k_0}$$

$$\text{(repeat the above method)}$$

$$\leq \|L_{k_0}\cdots L_m E^m L_m^T \cdots L_{k_0+1}^T H_{k_0}^T\| + 2M_e h_{k_0}$$

$$\leq \|L_{k_0}E^{k_0}H_{k_0}^T\| + 2M_e h_{k_0}$$

$$\text{(we use the definition: } E^k := E_{ss}^k := L_{k+1}E^{k+1}L_{k+1}^T \text{ for all } k)$$

$$= \|E_{sd}^{k_0-1}\| + 2M_e h_{k_0}$$

$$\text{(use the definition } E_{sd}^{k-1} := L_k E^k H_k^T, \text{ for all } k)$$

$$\leq 2M_e h_{k_0-1}. \qquad \text{(by (3.6.5) and } h_{k_0-1} = 2h_{k_0})$$

Hence (3.6.31) follows.

The proof of (3.6.32) is similar.

(3.6.33) is an immediate result of (3.6.30) and (3.6.31). $\square$

In each step of the algorithm we should ensure that the matrix $I - \tilde{E}^k - \tilde{F}^k$ is invertible. This is guaranteed by the following lemma.

Lemma 3.6.3 Suppose that $I - T$ has bounded inverse, and assume that for $k \geq m_0$, $I - E^k - F^k$ is invertible. Then for $m \geq 3m_0$, there exists k_0, such that for all $k : m > k \geq \max\left\{k_0, m_0, \left[\frac{m+2}{3}\right]\right\}$ we have

$$\|(I - \tilde{E}^k - \tilde{F}^k)^{-1}\| \leq M_c, \tag{3.6.34}$$

where M_c is a constant independent of k and m.

Proof. Under the condition of this lemma, for $k \geq m_0$ there exists a constant M such that

$$\|(I - E^k - F^k)^{-1}\| \leq M. \tag{3.6.35}$$

Therefore,

$$
\begin{aligned}
\|\tilde{E}^k - E^k\| &= \|\tilde{E}_{ss}^k + \tilde{E}_{sd}^k \tilde{\gamma}^k \tilde{E}_{ds}^k - E^k\| \quad \text{(from (3.2.17))} \\[6pt]
&\leq \|\tilde{E}_{ss}^k - E^k\| + \|\tilde{E}_{sd}^k\|\|\tilde{\gamma}^k\|\|\tilde{E}_{ds}^k\| \\[6pt]
&\leq \|\tilde{E}_{ss}^k - E^k\| + 2(2M_e h_k)^2 \quad \text{(from (3.6.31), (3.6.33))} \\[6pt]
&\leq \|L_{k+1}\tilde{E}^{k+1}L_{k+1}^T - E^k\| + 2(2M_e h_k)^2 \\
&\hspace{5em} \text{(from the fourth formula in (3.2.17))} \\[6pt]
&\leq \|L_{k+1}\tilde{E}_{ss}^{k+1}L_{k+1}^T - E^k\| + 2(2M_e h_{k+1})^2 + 2(2M_e h_k)^2 \\
&\hspace{5em} \text{(repeat the above procedure)} \\[6pt]
&\leq \|L_{k+1}\cdots L_m\tilde{E}^m L_m^T \cdots L_{k+1}^T - E^k\| + 4(2M_e h_k)^2 \\[6pt]
&= \|E^k - E^k\| + 4(2M_e h_k)^2 \quad \text{(since } \tilde{E}^m = E^m\text{)} \\[6pt]
&= 4(2M_e h_k)^2
\end{aligned}
$$

thus we have

$$\lim_{k \to \infty} \|E^k - \tilde{E}^k\| = 0. \tag{3.6.36}$$

Similarly we have

$$\lim_{k \to \infty} \|F^k - \tilde{F}^k\| = 0, \tag{3.6.37}$$

hence

$$\lim_{k \to \infty} \|(I - E^k - F^k) - (I - \tilde{E}^k - \tilde{F}^k)\| = 0. \tag{3.6.38}$$

Given any $\varepsilon_0 > 0, \varepsilon_0 M < 1$ there exists k_0, such that

$$\|(I - E^k - F^k) - (I - \tilde{E}^k - \tilde{F}^k)\| < \varepsilon_0 \qquad (3.6.39)$$

for all $k \geq k_0$. By (3.6.39), (3.6.35) and Lemma 3.5.2, we obtain

$$\|(I - \tilde{E}^k - \tilde{F}^k)^{-1}\| \leq \frac{M}{1 - \varepsilon_0 M}$$

for all $k \geq k_0$.   □

Lemma 3.6.4   Under the assumptions of Lemma 3.6.3, $r \geq 1$, and

$$k > \frac{\ln h}{\ln 2} \qquad (3.6.40)$$

then

$$\|\tilde{S}^k\| \leq M_g \qquad (3.6.41)$$

where M_g is a constant, independent of k.

Proof.   From (3.2.16) and the definition of $\tilde{S}^k$

$$\|\tilde{S}^k\| \leq \|\tilde{S}^k - \tilde{S}_k^{k+1}\| + \|\tilde{S}_k^{k+1}\|$$

$$= \|(I - \tilde{E}^k - \tilde{F}^k)^{-1}\rho^k\| + \|\tilde{S}_k^{k+1}\|$$

$$\leq M_c\|\rho^k\| + \|L_{k+1}\tilde{S}^{k+1}\| \qquad (\|\tilde{S}_k^{k+1}\| \leq \|\tilde{S}^{k+1}\|, \|\tilde{d}_k^{k+1}\| \leq \|\tilde{S}^{k+1}\|)$$

$$\leq [M_c(8M_e h_k M_f h_k^r + 8M_e h_k M_f h_k^r + 2M_f h_k^r) + 1]\|\tilde{S}^{k+1}\|$$

$$= [M_c(16M_e M_f h_k^{r+1} + 2M_f h_k^r) + 1]\|\tilde{S}^{k+1}\|$$

$$\leq \prod_{j=k}^{m-1} [1 + M_c(16M_e M_f h_j^{r+1} + 2M_f h_j^r)]\|\tilde{S}^m\|$$

$$\leq \prod_{j=k}^{m-1} [1 + M_d h_j^r] \|\tilde{S}^m\| \quad (M_d := M_c(16 M_e M_f h + 2 M_f))$$

$$\leq \left\{ \prod_{j=k}^{m-1} \exp(M_d h_j^r) \right\} \|S^m\|$$

$$\leq \left\{ \exp\left(\sum_{j=k}^{\infty} M_d h_j^r \right) \right\} \|(I - E^m - F^m)^{-1}\| \|g^m\| \quad \text{(see (3.1.11))}$$

$$\leq \left\{ \exp\left(M_d \cdot \frac{(2h)^r}{2^r - 1} \cdot h_k^r \right) \right\} M \|g\| \quad \text{(since } \|g^m\| \leq \|g\|, \text{ and (3.6.35))}$$

$$< \left\{ \exp\left(M_d \frac{(2h)^r}{2^r - 1} \right) \right\} M \|g\| \quad \text{(from (3.6.40), } h_k^r < 1)$$

$$= M_g.$$

$\square$

Lemma 3.6.5   Assume that k, m satisfy the conditions in Lemma 3.6.2, and (3.6.40). Then there exists a constant M_ρ such that

$$\|\rho^k\| \leq M_\rho h_k^r \tag{3.6.42}$$

where M_ρ is independent of k.

Proof.   By (3.2.19), (3.2.17) and Lemma 3.6.2

$$\|\rho^k\| = \|\tilde{E}_{sd}^k \tilde{\gamma}^k \tilde{F}_{ds}^k \tilde{S}_k^{k+1} + \tilde{E}_{sd}^k \tilde{\gamma}^k \tilde{F}_{dd}^k \tilde{d}_k^{k+1} + \tilde{F}_{sd}^k \tilde{d}_k^{k+1}\|$$

$$\leq 8 M_e M_f h_k^{r+1} \|\tilde{S}_k^{k+1}\| + 8 M_e M_f h_k^{r+1} \|\tilde{d}_k^{k+1}\| + 2 M_f h_k^r \|\tilde{d}_k^{k+1}\|$$

$$= 8 M_e M_f h_k^{r+1} \{ \|L_{k+1} \tilde{S}^{k+1}\| + \|H_{k+1} \tilde{S}^{k+1}\| \}$$

$$\quad + 2 M_f h_k^r \|H_{k+1} \tilde{S}^{k+1}\|$$

$$< 8 M_e M_f h_k^r \{ \|\tilde{S}^{k+1}\| + \|\tilde{S}^{k+1}\| \} + 2 M_f h_k^r \|\tilde{S}^{k+1}\| \quad \text{(by (3.6.40))}$$

$$\leq h_k^r \{ 16 M_e M_f M_g + 2 M_f M_g \} \quad \text{(by (3.6.41))}$$

$$= M_\rho h_k^r.$$

□

Remark From (3.6.42), (3.6.5) and (3.6.31), $\|\rho^k\|$ is of higher order comparing to $\|E_\&^k\|$ and $\|\tilde{E}_\&^k\|$, where $\& = sd, ds$ or dd.

Now we indicate an important result in this section.

Theorem 3.6.5 Let $r \geq 3s > 0, K > 1$, where r, s are the orders of smoothness for functions b and g respectively. If m and m_0 are positive integers such that $m > m_1 := \left[\frac{m+2}{3}\right] \geq m_0$, and $m \geq \max\{3C - 2, \frac{\ln h}{\ln 2}\}$, then

$$\|S^m - \overline{S}^m\| \leq M_a h_m^s. \tag{3.6.43}$$

where M_a is a constant independent of m, and $C = \dfrac{\ln(4M_e h)}{\ln 2}$.

Proof. By using (3.2.4), (3.3.2), and (3.1.15) we have

$$\|S^m - \overline{S}^m\| = \|L_m^T(S_{m-1}^m - \overline{S}^{m-1}) + H_m^T(d_{m-1}^m - \overline{d}^{m-1})\|$$

$$\leq \|S_{m-1}^m - \overline{S}^{m-1}\| + \|d_{m-1}^m - \overline{d}^{m-1}\|$$

$$\leq \|S_{m-1}^m - \overline{S}^{m-1}\| + \|\gamma^{m-1} E_{ds}^{m-1}\| \|S_{m-1}^m - \overline{S}^{m-1}\|$$

$$+ \|\gamma^{m-1} F_{ds}^{m-1}\| \|S_{m-1}^m\| + \|\gamma^{m-1} F_{dd}^{m-1} d_{m-1}^m\|.$$

From (3.6.5), (3.6.6), (3.2.17), (3.6.33) and (3.6.34) we obtain

$$\|S^m - \overline{S}^m\| \leq (1 + 2M_e h_{m-1}) \|S_{m-1}^m - \overline{S}^{m-1}\| + 4M_f M_c \|g\| h_{m-1}^r.$$

Denote $N_g := 4M_f M_c \|g\|$. From (3.2.5) and (3.2.9) we have

$$\|S^m - \overline{S}^m\| \leq (1 + 2M_e h_{m-1})(\|S_{m-1}^m - \tilde{S}^{m-1}\| + \|\tilde{S}^{m-1} - \overline{S}^{m-1}\|)$$

$$+ N_g h_{m-1}^r$$

$$= (1 + 2M_e h_{m-1})(\|\rho^{m-1}\| + \|\tilde{S}^{m-1} - \overline{S}^{m-1}\|) + N_g h_{m-1}^r.$$

By using (3.6.42) we obtain

$$\|S^m - \overline{S}^m\| \leq N_g h_{m-1}^r + (1 + 2M_e h_{m-1})\|\tilde{S}^{m-1} - \overline{S}^{m-1}\|$$

$$\leq N_g h_{m-1}^r + N_g \sum_{j=m_1}^{m-2} \left[\prod_{\nu=j+1}^{m-1} (1 + 2M_e h_\nu) \right] h_j^r$$

$$+ \left[\prod_{j=m_1}^{m-1} (1 + 2M_e h_j) \right] \cdot \|\tilde{S}^{m_1} - \overline{S}^{m_1}\|$$

$$\leq N_g h_{m-1}^r + N_g \sum_{j=m_1}^{m-2} \left[\exp \left(2M_e \sum_{\nu=j+1}^{m-2} h_\nu \right) \right] h_j^r$$

$$\leq N_g h_{m-1}^r + N_g \sum_{j=m_1}^{m-2} [\exp(4M_e h_{j+1})] h_j^r.$$

Since $M \geq 3C - 2$ and $m_1 \geq C - 1$, then

$$\exp(4M_e h_{m_1+1}) \leq 1.$$

We have

$$\|S^m - \overline{S}^m\| \leq N_g h_{m-1}^r + N_g \sum_{j=m_1}^{m-2} h_j^r$$

$$\leq N_g h_{m-1}^r + \frac{2^r N_g}{2^r - 1} h_{m_1}^r$$

$$\leq N_g \left(1 + \frac{2^r}{2^r - 1} \right) h_{m_1}^r,$$

where the last inequality is based on the fact that $r \geq 1$. Let

$$N_r := N_g \left(1 + \frac{2^r}{2^r - 1} \right),$$

we obtain

$$\|S^m - \overline{S}^m\| = N_r \left(\frac{2\pi}{K} \right)^{r - \frac{rm_1}{m}} (h_m)^{\frac{rm_1}{m}}$$

$$= N_r (2\pi)^{r - \frac{rm_1}{m}} K^{r(\frac{m_1}{m} - 1)} (h_m)^{\frac{rm_1}{m}}.$$

Since $K > 1$, $m_1 < m$, and $K^{r(\frac{m_1}{m}-1)} < 1$ we have

$$\|S^m - \overline{S}^m\| \; < (2\pi)^r N_r(h_m)^{\frac{rm_1}{m}}$$

$$= M_a(h_m)^{\frac{rm_1}{m}}$$

$$\leq M_a(h_m)^{\frac{3sm_1}{m}}$$

$$\leq M_a h_m^s.$$

The following is a crucial result for the error analysis.   □

Theorem 3.6.6   Under the conditions in Theorem 3.6.5, there exists a constant M, such that

$$\|u - \overline{u}_m\| \leq M h_m^s \tag{3.6.44}$$

where m is any integer satisfying the condition in Th.3.6.5.

Proof.   (3.1.3), (3.3.5), and (3.6.43) and (3.5.2) imply (3.6.44).   □

Now we discuss the condition (3.6.45) which appears in Theorem 3.5.1. We shall show that under the rest conditions in Theorem 3.5.1, the solution u of the integral equation $(3.0)_1$, is smooth, more precisely

$$u \in \overset{\circ}{C}{}^s ([0, 2\pi]) \tag{3.6.45}$$

which means the assumption (3.6.45) in Theorem 3.5.1 is redundant.

The Remark after Theorem 3.5.1 indicates that although in general the solutions of integral equation $(3.0)_1$ might not exist (we shall see this from the theory of integral equations which will be introduced below). But the case which we are concerned with is that the solution does exist and is unique.

Based on the fact indicated above we consider mainly the smoothness of the solution u. Before doing this we now quote some known results.

Definition 3.6.1   A complex-valued function $x(t)$ of a real variable t, defined in a finite interval $a \leq t \leq b$, and satisfying

$$\int_a^b |x(t)|^2 dt < \infty, \tag{3.6.46}$$

where the integral being taken in the sense of Lebesgue, will be called an L^2 function.

Let $K(s,t)$ be a function of two variables satisfying three conditions:

(a) $K(s,t)$ is a measurable function of (s,t) in the square $a \leq s \leq b, a \leq t \leq b$, such that

$$\int_a^b \int_a^b |K(s,t)|^2 ds dt < \infty; \tag{3.6.47}$$

(b) for each value of s, $K(s,t)$ is a measurable function of t such that

$$\int_a^b |K(s,t)|^2 dt < \infty; \tag{3.6.48}$$

(c) for each value of t, $K(s,t)$ is a measurable function of s such that

$$\int_a^b |K(s,t)|^2 ds < \infty. \tag{3.6.49}$$

A function $K(s,t)$ satisfying these conditions will be called an L^2 kernel.

We consider the linear integral equation of the second kind

$$x(s) = y(s) + \lambda \int_a^b K(s,t)x(t)dt, \tag{3.6.50}$$

where $K(s,t)$ is an L^2 kernel, $y(s)$ is an L^2 function, and λ is a complex parameter. We may abbreviate (3.6.50) as

$$x = y + \lambda Kx. \tag{3.6.51}$$

The related homogeneous linear integral equation is

$$x(s) = \lambda \int_a^b K(s,t)x(t)dt \quad (a \leq s \leq b) \tag{3.6.52}$$

or, in the abbreviated notation

$$x = \lambda Kx. \tag{3.6.53}$$

The equation (3.6.52) has the trivial solution $x(s) = 0$. If it has any L^2 solution $x(s)$ other than this trivial one, we call λ a **characteristic value** of $K(s,t)$, and the function $x(s)$ a **characteristic function** of $K(s,t)$ belonging to the characteristic value λ. The following Lemma 3.6.6 can be found in ([Sm, p.47 and 49]).

Lemma 3.6.6   Let $K(s,t)$ be an L^2 kernel. Then:
(a) either the equation

$$x(s) = y(s) + \lambda \int_a^b K(s,t)x(t)dt \qquad (3.6.54)$$

has a unique L^2 solution $x(s)$ for every L^2 function $y(s)$, or the associated homogeneous equation

$$x(s) = \lambda \int_a^b K(s,t)x(t)dt \qquad (3.6.55)$$

has an L^2 solution $x(s)$ that does not vanish identically;
(b) when λ is a characteristic value of K, equation (3.6.55) has at most a finite number of linearly independent L^2 solutions, and the equation (3.6.54) has an L^2 solution $x_0(s)$ for a given L^2 function $y(s)$. Therefore the general L^2 solution of (3.6.54) is given by

$$x(s) = x_0(s) + \sum_{\sigma=1}^p \alpha_\sigma x^{(\sigma)}(s), \qquad (3.6.56)$$

where $\{x^{(\sigma)}(s)\}$ is a full system of characteristic functions of $K(s,t)$ for the characteristic value λ.

Now we consider the integral equation $(3.0)_1$. Corresponding to (3.6.54), we take $a = 0, b = 2\pi$, and

$$K(x,y) = a(x-y)\ln\left|2\sin\frac{x-y}{2}\right| + b(x,y)$$

$$\qquad (3.6.57)$$

$$y(x) = g(x).$$

According to the assumption given in Theorem 3.5.1, $a, g \in \overset{\circ}{C}{}^s ([0, 2\pi]), b \in \overset{\circ}{C}{}^r$ $([0, 2\pi] \times [0, 2\pi]), r, s \le n$ and $r \ge 3s > 0$ (see Th. 3.6.5), by Minkowski inequality and

$$\int_0^{2\pi} \int_0^{2\pi} \ln^2|x - s| dx ds < \infty. \tag{3.6.58}$$

We conclude that $K(x, y)$ in (3.6.57) is an L^2 kernel.

Since we only consider the cases which the solutions of $(3.0)_1$ exist, from Lemma 3.6.6, the solutions are L^2 functions.

Since the L^2 solution u satisfies $(3.0)_1$, it must be a periodic function, i.e.,

$$u \in \overset{\circ}{L}_2 ([0, 2\pi]). \tag{3.6.59}$$

Before giving the proof of (3.6.45), we establish some lemmas.

Definition 3.6.2   A curve $\Gamma : z = z(\tau) = x(\tau) + iy(\tau), \tau \in [\alpha, \beta]$ is called regular if $z'(\tau) = x'(\tau) + iy'(\tau) \neq 0$ for all $\tau \in [\alpha, \beta]$ and the derivative $z'(\tau)$ is continuous.

The geometric interpretation of these properties is that a regular arc has a nonzero tangent at every point and it varies continuously along the curve.

Lemma 3.6.7   Let Γ be a regular arc and $\mu \in L_p(\Gamma), p > 1$, that is,

$$\int_\Gamma |\mu(t)|^p dt < \infty. \tag{3.6.60}$$

Then

$$u(z) = \int_\Gamma \mu(t) \ln \frac{1}{|t - z|} |dt| \tag{3.6.61}$$

is harmonic in $\mathbb{C}\backslash\Gamma$ and continuous on $\mathbb{C}$.

Proof.   That $u(z)$ is harmonic in $\mathbb{C}\backslash\Gamma$ is obvious. We now prove the second assertion. Since $u(z)$ is harmonic, it must be continuous when $z \overline{\in} \Gamma$. It remains to show for any $z_0 \in \Gamma$ there exists $\delta_0 > 0, \delta_0 = \delta(z_0)$ such that for all z satisfying $|z - z_0| < \delta_0$, there holds

$$|u(z) - u(z_0)| < \varepsilon. \tag{3.6.62}$$

Since Γ is regular, for sufficiently small $\rho > 0$ the part of Γ within the disc $|z - z_0| \le \rho$ is a single arc. Denote this arc by γ_1 and the rest by γ_2. We write

$$f(z) = u(z) - u(z_0)$$

$$= \left(\int_{\gamma_1} + \int_{\gamma_2} \right) \mu(t) \ln \left| \frac{z_0 - t}{z - t} \right| d\tau \qquad (3.6.63)$$

$$= f_1(z) + f_2(z).$$

Let q be such that $q^{-1} + p^{-1} = 1$, then by Hölder's inequality,

$$|f_1(z)| \le \|\mu\|_p \left\{ \int_{\gamma_1} \ln \left| \frac{z_0 - t}{z - t} \right|^q d\tau \right\}^{\frac{1}{q}}. \qquad (3.6.64)$$

By assumption $\mu \in L_p(\Gamma)$, thus the first factor on the right is bounded. As for the second factor, since

$$(a + b)^q \le 2^q (a^q + b^q) \quad \text{for all} \quad b \ge 0, a \ge 0,$$

if we let $a = |\ln|z_0 - t||$ and $b = |\ln|z - t||$, then

$$|\ln|z_0 - t| - \ln|z - t||^q \le 2^q \left[\left| \ln \frac{1}{|z_0 - t|} \right|^q + \left| \ln \frac{1}{|z - t|} \right|^q \right].$$

The second factor on the right side of (3.6.64) is less than

$$2 \left\{ \int_{\gamma_1} \left| \ln \frac{1}{|z_0 - t|} \right|^q d\tau + \int_{\gamma_2} \left| \ln \frac{1}{|t_z - t|} \right|^q d\tau \right\}^{\frac{1}{q}} \qquad (3.6.65)$$

where t_z is the point on γ_1 closest to z. The two integrals in (3.6.65) tend to zero as ρ tends to zero. Given $\varepsilon > 0$ there exists ρ such that

$$|f_1(z)| < \frac{1}{2}\varepsilon \qquad \text{for all} \quad z. \qquad (3.6.66)$$

Since $\gamma_2 = \Gamma \backslash \gamma_1$, $\ln \left| \frac{z_0 - t}{z - t} \right|$ tends to zero if $z \to z_0$ for $t \in \gamma_2$. Thus there exists $\delta > 0$ such that

$$|f_2(z)| < \frac{1}{2}\varepsilon, \quad \text{for} \quad |z - z_0| < \delta. \qquad (3.6.67)$$

From (3.6.66), (3.6.67) and (3.6.63) we have (3.6.62). Thus we have completed the proof of the second assertion. $\qquad \square$

Corollary 3.6.1   Under the hypothesis on a, b and g in Theorem 3.5.1, the solution $u(x)$ of $(3.0)_1$ is of $H(1 - \varepsilon)$ (Hölder's condition with Hölder index $1 - \varepsilon$), where ε is any positive number less than 1.

Proof.   Since $u(x)$ is a solution of $(3.0)_1$, it satisfies the following equation

$$u(x) = \int_0^{2\pi} u(y) \left\{ a(x - y)\ln \left| 2 \sin \frac{x - y}{2} \right| + b(x, y) \right\} dy + g(x). \qquad (3.6.68)$$

Set

$$I(x) = \int_0^{2\pi} u(y) a(x - y)\ln \left| 2 \sin \frac{x - y}{2} \right| dy$$

$$= \int_0^{2\pi} u(y) a(x - y)\ln \left| \frac{2 \sin \dfrac{x - y}{2}}{x - y} \right| dy + \int_0^{2\pi} u(y) a(x - y)\ln |x - y| dy$$

$$= I_1(x) + I_2(x).$$

$$(3.6.69)$$

The first integral $I_1(x)$ in (3.6.69) is continuous since $a \in \overset{\circ}{C}{}^s ([0, 2\pi])$. On the other hand, $u \in \overset{\circ}{L}_2 ([0, 2\pi])$, hence $I_2(x)$ is also continuous by Lemma 3.6.7, thus $I(x)$ is continuous. Since $b \in \overset{\circ}{C}{}^r ([0, 2\pi] \times [0, 2\pi])$ and $g \in \overset{\circ}{C}{}^s ([0, 2\pi])$, by above discussion, each term in the right of (3.6.68) is continuous, hence $u(x)$ is continuous.

Since $a \in \overset{\circ}{C}{}^s ([0, 2\pi])$, $s > 0$, a belongs to $H(1)$, and hence $a(t)\ln \left| 2 \sin \frac{t}{2} \right| \in H(1 - \varepsilon)$, any $0 < \varepsilon < 1$, (cf. p.160 Note 3) and

$$|I(x_1) - I(x_2)| \leq \int_0^{2\pi} |u(y)| \left| a(x_2 - y)\ln \left| 2 \sin \frac{x_1 - y}{2} \right| \right.$$

$$\left. - a(x_2 - y)\ln \left| 2 \sin \frac{x_2 - y}{2} \right| \right| dy$$

$$\leq \left(\max_{0 \leq y \leq 2\pi} |u(y)| \right) \cdot C |x_1 - x_2|^{1-\varepsilon} \qquad (3.6.69)_1$$

by $(3.6.69)_1$ and the smoothness of b and g. The proof is completed.   □

Corollary 3.6.2   Under the hypothesis of a, b and g in Theorem 3.5.1, the solution $u(x)$ of $(3.0)_1$ is of $\overset{\circ}{C}{}^1([0, 2\pi])$.

Proof.   Since $a, g \in \overset{\circ}{C}{}^s([0, 2\pi])$ and $b \in \overset{\circ}{C}{}^r([0, 2\pi] \times [0, 2\pi])$, we only need to prove $I(x) \in C^1([0, 2\pi])$, where $I(x)$ is defined in (3.6.69), $I(x) = I_1(x) + I_2(x)$.

$$I_1(x) = \int_0^{2\pi} u(y)a(x-y)\ln\left|\frac{2\sin\frac{x-y}{2}}{x-y}\right| dy \qquad (3.6.70)$$

Since

$$\left|\frac{2\sin\frac{x-y}{2}}{x-y}\right| = \left|\frac{e^{i(x-y)}-1}{x-y}\right| = \left|\sum_{n=1}^{\infty}\frac{i^n}{n!}(x-y)^{n-1}\right|, \qquad (3.6.71)$$

for any fixed y, from (3.6.71), it is clear that the function

$$\frac{2\sin\frac{x-y}{2}}{x-y}$$

is differentiable and belongs to $C^{\infty}([0, 2\pi])$; in fact, for any positive integer r

$$\left[\sum_{n=1}^{\infty}\frac{i^n}{n!}(x-y)^{n-1}\right]^{(r)} = \sum_{\nu=0}^{\infty}\frac{i^{\nu+1+r}}{\nu+r+1}\frac{(x-y)^{\nu}}{\nu!} \qquad (3.6.72)$$

the absolute value of the right side of (3.6.72) is less than $e^{(x-y)}$. Therefore

$$I_1(x) \in C^s([0, 2\pi]). \qquad (3.6.73)$$

We now consider $I_2(x)$,

$$I_2(x) = \int_0^{2\pi} u(y)a(x-y)\ln|x-y|dy \qquad (3.6.74)$$

From Corollary 3.6.1 and $a \in \overset{\circ}{C}{}^s([0, 2\pi])$, we conclude that $u(\cdot)a(x-\cdot) \in H(1-\varepsilon)$.

The derivative of $I_2(x)$ is

$$\frac{dI_2(x)}{dx} = \int_0^{2\pi} u(y)a'(x-y)\ln|x-y|dy + \int_0^{2\pi}\frac{u(y)a(x-y)}{x-y}dy$$

$$= F_1(x) + F_2(x) \qquad (3.6.75)$$

where $F_1(x) = \int_0^{2\pi} u(y)a'(x-y)\ln|x-y|dy$. By Lemma 3.6.7 $F_1(x)$ is a continuous function; and

$$F_2(x) = \int_0^{2\pi} \frac{u(y)a(x-y)}{x-y}dy, \quad \text{if} \quad x \neq 0 \quad \text{and} \quad 2\pi. \tag{3.6.76}$$

The integral on the right side of (3.6.75) takes the principal value. By the Sokhotskyi–Plemelj formula,

$$F_2(x) = -a(0)u(x)[\pi i + \ln(x-2\pi) - \ln x]$$

$$+ \int_0^{2\pi} \frac{u(y)a(x-y) - a(0)u(x)}{x-y}dy \tag{3.6.77}$$

(cf. p.160 Note 4).

Since $u(\cdot)a(x - \cdot) \in H(1-\varepsilon)$, then F_2 also belongs to $H(1-\varepsilon)$ (cf. p.162 Note 5).

We conclude that $I_2(x)$ is differentiable and that $dI_2(x)/dx$ is continuous at every point $x \in (0, 2\pi)$.

If $x = 0$ we can alter the interval of integration from $[0, 2\pi]$ to $[-\pi, \pi]$ since the integrand in $I(x)$ is

$$u(y)a(x-y) \cdot \ln\left|2\sin\frac{x-y}{2}\right|$$

it is a periodic function with respect to variable y. Thus $x = 0$ is an inner point of $[-\pi, \pi]$, and we can still use the Sokhotskyi–Plemelj formula and have the same conclusion. The proof of Corollary 3.6.2 is completed.

□

In the following we indicate a result on the order of smoothness of $u(x)$.

Theorem 3.6.7   Assume that $a, g \in \overset{\circ}{C}{}^s$ $([0, 2\pi]), b \in \overset{\circ}{C}{}^r$ $([0, 2\pi] \times [0, 2\pi]), 1 \leq s \leq r$. Then the solution $u(x)$ of the integral equation $(3.0)_1$, if exists, belongs to $\overset{\circ}{C}{}^s$ $([0, 2\pi])$ (cf. p.163 Note 9).

Proof.   From the discussion in the proof of Corollary 3.6.2, without loss of generality we may assume that $x \neq 0, 2\pi$, i.e., $x \in (0, 2\pi)$.

If $s = 1$, then the assertion follows from Corollary 3.6.2.

If $s > 1$, we prove by induction.

We first prove the derivative of $u(x)$, $u^{(1)}(x)$ belongs to $H(\nu)$ for some $0 < \nu \leq 1$.

Since $s \geq 2, g, a \in \overset{\circ}{C}{}^s, b \in \overset{\circ}{C}{}^r, r \geq s$, from (3.6.68), (3.6.69), (3.6.73), (3.6.74), (3.6.75) and (3.6.77), we only need to check whether the functions $F_1(x)$ and $F_2(x)$ satisfy the $H(\nu)$ condition for some ν.

Since $u \in \overset{\circ}{C}{}^1$ by Corollary 3.6.2, $u(\cdot)a^{(1)}(x - \cdot) \in \overset{\circ}{C}{}^1$ and $F_1(x)$ is differentiable:

$$F_1^{(1)}(x) = \int_0^{2\pi} u(y) \left[a^{(2)}(x - y)\ln|x - y| + a^{(1)}(x - y)\frac{1}{x - y} \right] dy$$

where the principal value must be taken (cf. *p.162 Note 6*).

On the other hand, since

$$F_2(x) = \int_0^{2\pi} \frac{u(y)a(x - y)}{x - y} dy$$

belongs to $H(1 - \varepsilon)$, from (3.6.75)

$$\frac{dI_2}{dx} = F_1 + F_2 \in H(1 - \varepsilon);$$

from (3.6.73), (3.6.68) and (3.6.69),

$$u^{(1)} \in H(1 - \varepsilon), \quad 0 < \varepsilon < 1. \tag{3.6.78}$$

Suppose that $u^{(k)} \in H(\nu)$ for some $\nu, 0 < \nu \leq 1$, and $k + 1 < s$. We prove below that $u^{(k+1)}$ exists and belongs to $H(\mu), 0 < \mu \leq 1$.

Since $u^{(k+1)}$ can be written formally (refer to (3.6.68)) into

$$u^{(k+1)}(x) = \phi_1^{(k+1)}(x) + \phi_2^{(k+1)}(x) + g^{(k+1)}(x), \tag{3.6.79}$$

where

$$\phi_1(x) = \int_0^{2\pi} u(y)m(x, y)dy \tag{3.6.80}$$

$$\phi_2(x) = \int_0^{2\pi} u(y)b(x, y)dy \tag{3.6.81}$$

$$m(x, y) = a(x - y)\ln \left| 2\sin\frac{x - y}{2} \right|. \tag{3.6.82}$$

Since $g \in \overset{\circ}{C}{}^{s}, b \in \overset{\circ}{C}{}^{r}, k + 1 \leq s \leq r$. The derivatives $g^{(k+1)}$ and $\phi_2^{(k+1)}$ do exist. We have

$$\phi_1^{(k)}(x) = \int_0^{2\pi} m(x, y) u^{(k)}(y) dy. \tag{3.6.83}$$

Since $u^{(k)} \in H(\nu)$, the integral in (3.6.83) exists in the sense of principal value, hence $\phi_1^{(k)}(x)$ is differentiable (cf. p.162 Note 7). i.e., $\phi_1^{(k+1)}(x)$ exists.

Now if $k = s - 1$ (cf. p.162 Note 8), define

$$V_\epsilon(x) = \int_0^{x-\epsilon} m(x, y) u^{(s-1)}(y) dy + \int_{x+\epsilon}^{2\pi} m(x, y) u^{(s-1)}(y) dy.$$

Then

$$\frac{dV_\epsilon(x)}{dx}$$

$$= \left[a(x - y) \ln \left| 2 \sin \frac{x-y}{2} \right| \right]_{y=x+\epsilon}^{y=x-\epsilon}$$

$$+ \int_0^{x-\epsilon} + \int_{x+\epsilon}^{2\pi} \left(\frac{\partial}{\partial x} m(x, y) \right) u^{(s-1)}(y) dy$$

$$= [a(\epsilon) - a(-\epsilon)] \ln \left| 2 \sin \frac{\epsilon}{2} \right| + \int_0^{x-\epsilon}$$

$$+ \int_{x+\epsilon}^{2\pi} u^{(s-1)}(y) \left[\frac{\partial a(x-y)}{\partial x} \cdot \ln \left| 2 \sin \frac{x-y}{2} \right| + \frac{a(x-y)}{2} \cot \frac{x-y}{2} \right] dy,$$

and it is evident that $\lim\limits_{\epsilon \to 0} \dfrac{dV_\epsilon(x)}{dx}$ exists, i.e.,

$$\phi_1^{(s)}(x) = \int_0^{2\pi} u^{(s-1)}(y) \left[\frac{\partial a(x-y)}{\partial x} \ln \left| 2 \sin \frac{x-y}{2} \right| \right.$$

$$\left. + \frac{a(x-y)}{2} \cot \frac{x-y}{2} \right] dy \tag{3.6.84}$$

the integral in (3.6.84) is taken as the principal value (cf. p.149 Note 4). It is easy to prove (3.6.83) is true for $k = 1$, thus we complete the proof. $\square$

By Theorem 3.6.7 and Theorem 3.5.1 we have (3.5.2), and then (3.6.44) is valid.

§3.7   The Dirichlet Problem

The method introduced above can be applied to solve the Dirichlet problem.

Let D be a planar domain. For simplicity, we assume that D is a bounded, simply connected domain.

A function u which satisfies the Laplace equation

$$\Delta u(z) = 0, \quad z \in D \tag{3.7.1}$$

and boundary condition

$$u(z) = f(z), \quad z \in \Gamma \tag{3.7.2}$$

where $\Gamma = \partial D$, is called a solution of the Dirichlet problem, and where f is a given function and z ($= x + iy$) is a complex variable on the complex plane $\mathbb{C}$, on occasion we also use a pair of two real variables (x, y) instead of z.

The following Theorem 3.7.1 converts the Dirichlet problem into a problem of solving an integral equation of type $(3.0)_1$.

Theorem 3.7.1   Let D be a simply connected domain, $\partial D = \Gamma$, Γ an analytic Jordan closed curve, with a parametric representation

$$\Gamma : z(t) = x(t) + iy(t), \quad 0 \le t \le 2\pi, \tag{3.7.3}$$

satisfying

$$0 < |\dot{z}(t)| < \infty, \quad 0 \le t \le 2\pi. \tag{3.7.4}$$

Let $\phi(z)$ be the solution of the following integral equation

$$\phi(z) = -2 \int_\Gamma \phi(w) \left[\frac{\partial \Phi(z, w)}{\partial n(w)} + \alpha \Phi(z, w) \right] ds(w) + 2f(z) \tag{3.7.5}$$

or, equivalently, $\tilde{\phi}(t)$ ($\tilde{\phi}(t) := \phi(z(t))$) satisfies

$$\tilde{\phi}(t) + \int_0^{2\pi} \tilde{\phi}(\tau) \left[a(t, \tau)\ln \left| 2\sin \frac{t - \tau}{2} \right| + b(t, \tau) \right] d\tau = \tilde{f}(t) \quad 0 \le t \le 2\pi \tag{3.7.6}$$

where

$$\Phi(z,w) = \frac{1}{2\pi} \ln \frac{1}{|z-w|},$$

$n(w)$ is a normal vector at point $w \in \Gamma$ directed into D. α is a constant,

$$\tilde{f}(t) = 2f(z(t)), \quad a(t,\tau) = -\frac{\alpha}{\pi}, \quad w = z(\tau), \quad z = z(t),$$

and

$$b(t,\tau) = 2\frac{\partial \Phi(z(t), z(\tau))}{\partial n(z(\tau))} + \frac{\alpha}{\pi} \ln \left| \frac{2 \sin \frac{t-\tau}{2}}{t-\tau} \right|$$

$$+ \frac{\alpha}{\pi} \ln \left| \frac{t-\tau}{z(t) - z(\tau)} \right|. \tag{3.7.7}$$

Then the Dirichlet problem (3.7.1)–(3.7.2) is solved, the solution is

$$u(z) = \int_\Gamma \phi(w) \left[\frac{\partial \Phi(z,w)}{\partial n(w)} + \alpha \Phi(z,w) \right] ds(w) \tag{3.7.8}$$

where $s(w)$ is the arc length at the point $w (\in \Gamma)$ counted from some point on Γ. Moreover

$$u \in C^s([0, 2\pi]) \quad \text{if} \quad f \in C^s([0, 2\pi]). \tag{3.7.8}'$$

Proof.   From the definition of function $\Phi(z, w)$ and the right side of (3.7.8), it is clear that $u(z)$ is harmonic on D. By using the SP theorem (see Note 4), we have

$$u(z_0) = \int_\Gamma \phi(w) \left[\frac{\partial \Phi(z_0, w)}{\partial n(w)} + \alpha \Phi(z_0, w) \right] ds(w) + \frac{1}{2}\phi(z_0). \tag{3.7.9}$$

Substituting the value $\phi(z)$ of the right side of (3.7.5) into (3.7.9) we have

$$u(z_0) = f(z_0) \quad z_0 \in \Gamma$$

and conclude that $u(z)$ as defined by (3.7.8) is the solution of the problem (3.7.1)–(3.7.2).

Note $a(t, \tau)$ is the constant $-\alpha/\pi$. We now consider $b(t, \tau)$, which consists of three terms, and

$$\frac{\partial \Phi(z, w)}{\partial n(w)} \cdot 2\pi = \frac{\partial}{\partial n(w)}(\ln|w - z|^{-1})$$

$$= \frac{\partial}{\partial n}(\ln \gamma^{-1})$$

$$= -\frac{1}{r}\frac{\partial r}{\partial n}$$

$$= \frac{d\omega}{ds}, \qquad (3.7.10)$$

where $r = |w - z| = |z(\tau) - z(t)|$, $w - z = re^{i\omega}$. Let $z(t) = x(t) + iy(t)$ and $z(\tau) = x(\tau) + iy(\tau)$ be respectively the fixed and variable points on Γ. Set $\omega = \arctan\dfrac{y(\tau) - y(t)}{x(\tau) - x(t)}$, then from (3.7.10)

$$2\pi\frac{\partial \Phi(z, w)}{\partial n(w)} = \frac{d\omega}{ds} = \begin{cases} \dfrac{y'(\tau)[x(\tau) - x(t)] - x'(\tau)[y(\tau) - y(t)]}{[x(\tau) - x(t)]^2 + [y(\tau) - y(t)]^2}, & \tau \neq t \\[3mm] \dfrac{1}{2}\dfrac{\dot{y}(t)\,\ddot{x}(t) - \dot{x}(t)\,\ddot{y}(t)}{[\dot{x}(t)]^2 + [\dot{y}(t)]^2}, & \tau = t \end{cases}$$

$$(3.7.11)$$

from the assumption (3.7.4) and (3.7.11), $d\omega/ds$ is finite, and belongs to C^∞ since Γ is analytic.

The function $\ln\left|\dfrac{2\sin\frac{1}{2}(t-\tau)}{t-\tau}\right|$ is also in C^∞, in fact, we have

$$\frac{d^r}{dt^r}\left(\frac{2\sin\dfrac{1}{2}(t - \tau)}{t - \tau}\right)\Bigg|_{t=\tau} = \begin{cases} 0, & r \text{ odd} \\[3mm] \dfrac{1}{r+1}, & r \text{ even} \end{cases} \qquad (3.7.12)$$

and it is differentiable at any point $\tau \in [0, 2\pi]$.

The function $\ln\left|\dfrac{t-\tau}{z(t)-z(\tau)}\right|$ is also in C^∞, since Γ is a Jordan curve, analytic, and satisfying the condition (3.7.4). In fact

$$\frac{d}{dt}\ln\left|\frac{t - \tau}{z(t) - z(\tau)}\right| = \begin{cases} \dfrac{z(t) - z(\tau) - (t - \tau)\dot{z}(t)}{[z(t) - z(\tau)](t - \tau)}, & t \neq \tau \\[3mm] -\dfrac{\ddot{z}(\tau)}{2\dot{x}(\tau)}, & t = \tau. \end{cases} \qquad (3.7.13)$$

By (3.7.11)–(3.7.13), $b(t, \tau) \in C^{\infty}([0, \pi]) \times C^{\infty}([0, 2\pi])$. From (3.7.5), it is clear that

$$\phi \in C^s([0, 2\pi]) \quad \text{if} \quad f \in C^s([0, 2\pi]). \tag{3.7.14}$$

By (3.7.8), (3.7.14), $u \in C^{(s)}([0, 2\pi])$, hence (3.7.8)' is proved.

It is easy to check the equivalence of (3.7.5) and (3.7.6).   □

The exterior Dirichlet problem for the Laplace equation

$$\Delta u = 0, \quad \text{for} \quad z = (x, y) \in R^2 \backslash D$$

$$u(z) = f(z) \quad \text{for} \quad z \in \Gamma (= \partial D) \tag{3.7.15}$$

can also be solved by solving an analogous integral equation ([KS1], [Ya2]),

As for the application of the integral equation $(3.0)_1$, one refers to Note 1.

Note 1   It is well known that the integral equation $(3.0)_1$ has a strong mechanical and physical background. In particular, the exterior boundary problem of acoustic wave scattering can be reduced to such an integral equation. Since the conformal mapping can be obtained from solving the related Dirichlet problem, and the latter can be turned into this kind of integral equation, therefore a wide range problems can be covered; we list some of them:

diffraction of electromagnetic waves;

atomic physics (the theory of interacting collisions);

nonlinear diffusion problems;

solidification problems;

theory of elasticity as applied to anisotropic media ; determination of cut-off frequencies of electromagnetic waveguides (also applications to mathematically related fields such as acoustical waveguide theory and membrane vibrations);

ion optics;

propagation of optical modes in dielectric fibers;

radio propagation in the atmosphere;

flow and heat transfer in ducts of arbitrary shape;

unsteady heat conduction problems (also diffusion processes following Fick's law);

supersonic flows;

plate theory (flexure, stability, and vibrations problems); etc. (refer to [SL]).

Note 2   The main parts of this chapter are taken from Chen-Peng's work [CP2], where the complexity order for the complexity order for the stiffness matrix is also estimated, and the method in the proof is changed in some extent.

Note 3   Since $a \in \overset{\circ}{C}{}^s$, $a \in H(1)$. We now prove $a(t)\ln\left|2\sin\frac{t}{2}\right| \in H(1-\varepsilon)$. In fact,

$$a(t)\ln\left|2\sin\frac{t}{2}\right| = a(t)\left[\ln\left|\frac{2\sin\frac{t}{2}}{t}\right| + \ln|t|\right]$$

$$= I_1 + I_2,$$

$$I_1 = a(t)\ln\left|\frac{2\sin\frac{t}{2}}{t}\right| \in H(1) \quad \text{since} \quad a \in \overset{\circ}{C}{}^s,$$

$$I_2 = a(t)\ln|t| = (a(t)t^{1-\varepsilon})(t^\varepsilon\ln|t|) \le M|t|^{1-\varepsilon}$$

where

$$M = \max_{t\in[0,2\pi]}(a(t)t^{1-\varepsilon}),$$

for any $\varepsilon, 0 < \varepsilon < 1$. It is easy to check $I_2 \in H(1-\varepsilon)$. Therefore, $a(t)\ln\left|2\sin\frac{t}{2}\right| \in H(1-\varepsilon)$.

Note 4   Sokhotsky–Plemelj (SP) formula. In the books [LS] and [H] the authors call this the Sokhosky formula, but in some other books, for instance in [Mu], the author calls it the Plemelj' formula. Here we call this the Sokhotsky-Plemelj' formula, because this formula was first proved by Sokhotsky in 1873, and proved again by Plemelj' in 1908.

We explain some preliminary concepts.

Regular arc   If the arc $\Gamma : z(\tau) = x(\tau) + iy(\tau)$ has non-zero tangent at each point, i.e., $\dot{z}(\tau) = \dot{x}(\tau) + i\dot{y}(\tau) \ne 0$, Γ is called a regular arc.

Cauchy integral   Let Γ be a simple regular curve, not necessarily closed, h be a complex-valued function defined on Γ. Let $z \neq \Gamma$. The function

$$f(z) := \frac{1}{2\pi i} \int_{\Gamma} \frac{h(\zeta)}{\zeta - z} d\zeta \qquad (*)$$

is known as the Cauchy integral.

Principal value (PV)   Let z_0 be a point on Γ which, if Γ is not closed, is not an endpoint. For $\rho > 0$ sufficiently small. Γ will intersect the circle $\Delta : |z - z_0| = \rho$ by Γ_ρ the portion of Γ lying in the circle Δ. The integral

$$\frac{1}{2\pi i} \int_{\Gamma - \Gamma_\rho} \frac{h(t)}{t - z} dt$$

exists for any $\rho > 0$. If its limit for $\rho \to 0$ also exists, the limit is called the principal value of the integral $(*)$, for $z = z_0$. By an analogous calculation we can obtain (3.6.77)—the Principal value of (3.6.76).

Left side and right side of Γ.   Let $\Gamma : z = z(\tau), \alpha \leq \tau \leq \beta$, be a regular arc, and let $z_0 := z(\tau_0), \alpha < \tau_0 < \beta$. Then the complex number $\dot{z}(\tau_0)$ represents a tangent vector of Γ at z_0 which points in the direction of increasing parameters. The complex number $i z(\tau_0)$ represents a vector perpendicular to the tangent vector which points to what we shall call the left side of Γ, the complex number $-i\dot{z}(\tau_0)$ points to the right side of Γ. If D is a sufficiently small disk centered at z_0, the set $D \cap \Gamma$ will have precisely two components. One component, which we call D^+, contains the points $z_0 + iz(\tau_0)\sigma$, the other, D^-, contains the points $z_0 - i\dot{z}(\tau_0)\sigma$ for sufficiently small $\sigma > 0$.

$f^+(z_0), f^-(z_0)$.   Let f be a function defined on Γ, $z_0 \in \Gamma$, we write

$$f^+(z_0) = \lim_{\substack{z \to z_0 \\ z \in D^+}} f(z),$$

$$f^-(z_0) = \lim_{\substack{z \to z_0 \\ z \in D^-}} f(z),$$

provided that the limits involved exist.

These will be called the unrestricted one-sided limits of f at $z_0 \in \Gamma$ from the left and from the right, respectively.

SP Theorem   Let γ be a simple regular curve, and let φ be a complex-valued function defined on γ which belongs to $H(\mu)$ (i.e., Lipμ, $0 < \mu < 1$). Let Γ be a curve such that:

(i) $\Gamma = \gamma$, if γ is closed.

or

(ii) Γ is any subarc of γ whose endpoints are interior points of γ, if γ is not closed.

Then the restriction of the principal value of the Cauchy integral (∗) to Γ likewise belongs to $H(\mu)$. Moreover, the one-side limits f^+ and f^- exist uniformly for $z_0 \in \Gamma$ without restriction on the approach. The SP formulas hold in the form

$$f^+(z_0) = f(z_0) + \frac{\varphi(z_0)}{2}$$

$$f^-(z_0) = f(z_0) - \frac{\varphi(z_0)}{2}$$

and the functions f^+ and f^- belong to $H(\mu)$ to Γ.

As for the validity of (3.6.75), i.e.,the differentiability of function

$$V(x) := \int_0^{2\pi} \varphi(y)\ln|x - y|dy,$$

where $\varphi \in H(\nu)$, one may refer to [Mu, page 31].

Note 5   The fact that

$$F_2(x) = \int_0^{2\pi} \frac{u(y)a(x - y)}{x - y}dy$$

belongs to $H(1 - \varepsilon)$ if $u(\cdot)a(x - \cdot) \in H(1 - \varepsilon)$ follows directly from the SP formula.

Note 6   Since $u(\cdot)a^{(1)}(x - \cdot) \in \overset{\circ}{C}^1$, the differentiability of F_1 (see (3.6.75)) follows from formula (13.3) in [Mu, page 31].

Note 7   The reason is the same as in Note 6.

Note 8   Since

$$u^{(s-1)}(x) = \phi_1^{(s-1)}(x) + \phi_2^{(s-1)}(x) + g^{(s-1)}(x)$$

we only need to show $\phi_1^{(s-1)}(x)$ is differentiable. But this can be proved by an approach analogous to that in the proof of Corollary 3.6.1.

Note 9   This result of Theorem 3.6.7 leads us to assure that an assumption in Theorem 3.5.1 ($u \in \overset{\circ}{C}{}^{s}([0, 2\pi])$) can be omitted.

After finishing the manuscript of this book, we introduced this part of the book to Professor Yuesheng Xu, he indicated that the smoothness of the solution u was already proved by Yunhe Zhao and him, and was presented in their paper [XZ].

Note 10   As was indicated in Theorem 3.4.1, where the complexity was estimated such that the stiffness matrix was calculated beforehand; it means that only the complexity in solving the linear equation was estimated.

Chapter 4

THE PERIODIC CARDINAL
INTERPOLATORY
WAVELETS

Periodic problems appear in various physical phenomena and mathematics which motivate an extensive study of periodic multi-resolution analysis (see [Me], [NW], [PB], [PT1], [PT2], [CM], [PT3], [C2], [C8] , [C9], [C10], [CLJ], [CLP] and [CP1]).

The scaling functions and wavelets (or quasi-wavelets) constructed in these works are mainly concerned with the number of terms in decomposition and reconstruction formulas, orthogonality, and regularity. These are very important in some mathematical and physical problems, as was shown in Chapter III. But in some other problems, more special properties are needed. The study of localization has been found only in [NW].

In the following we shall introduce a kind of periodic multi-resolution analysis in which the scaling function and wavelet possess the comprehensive properties: interpolation; localization; explicit analytic representations; symmetry; any order of regularity; real valued and bi-orthogonality (cf. [CLP], [CP3] & [CX]).

§4.1 The Periodic Cardinal Interpolatory Scaling Functions (PISF)

Before we construct the PISF we shall first give the definition of the generators for periodic multi-resolution analysis.

Let j be a nonnegative integer, K a positive integer, and $K_j = 2^j K, h_j = 2\pi/K_j$. Let $g(x)$ be a 2π-periodic, continuous differentiable function whose

164

Fourier coefficients are positive, i.e.,

$$g(x) = \sum_{n \in \mathbf{Z}} d_n e^{inx}, \quad d_n > 0, \quad \text{for all} \quad n \in \mathbf{Z} \tag{4.1.1}$$

Define

$$V_j = \text{span}\{g(\cdot), g(\cdot - h_j), \cdots, g(\cdot - (K_j - 1)h_j)\}, \tag{4.1.2}$$

then $\dim V_j = K_j$ and $V_j \subset V_{j+1}$, (see p.211 Note 1).

Definition 4.1.1 For $l = 0, 1, \cdots, K_j - 1$, define

$$Z_l^j(x) := \sum_{k=0}^{K_j-1} g(x + kh_j) e^{iklh_j} \tag{4.1.3}$$

$$= K_j \sum_{n \in \mathbf{Z}} d_{nK_j - l} e^{i(nK_j - l)x}, \tag{4.1.4}$$

$$\tilde{Z}_l^j(x) := \frac{Z_l^j(x)}{\|Z_l^j\|}. \tag{4.1.5}$$

It is easy to check that

$$\langle \tilde{Z}_{l_1}^j, \tilde{Z}_{l_2}^j \rangle = \delta_{l_1, l_2}, \quad 0 \le l_1, l_2 \le K_j - 1, \tag{4.1.6}$$

$$Z_l^j(x + kh_j) = e^{-ilkh_j} Z_l^j(x). \tag{4.1.7}$$

Since $Z_l^j(0) > 0$ we can give the following definition:

Definition 4.1.2 The following function is called a periodic cardinal interpolatory scaling function (PISF).

$$\phi_j(x) := \frac{1}{K_j} \sum_{l=0}^{K_j-1} \frac{Z_l^j(x)}{Z_l^j(0)}, \quad 0 \le j, \quad j \in \mathbf{Z}. \tag{4.1.8}$$

The following theorem presents the basic properties of the function ϕ_j.

Theorem 4.1.1 Suppose g is defined as in (4.1.1), V_j and ϕ as in (4.1.2) and (4.1.8) respectively. Then

(1) $\phi_j(x)$ possesses the cardinal interpolatory property, i.e.,

$$\phi_j(kh_j) = \delta_{0,k} \ \text{ for } \ k = 0, 1, \cdots, K_j - 1.$$

(2) $\{\phi_j(\cdot - kh_j)\}_{k=0}^{K_j-1}$ is a basis for V_j and

$$V_j = \text{span}\{\phi_j(\cdot - kh_j) : k = 0, 1, \cdots, K_j - 1\}.$$

(3) $\phi_j(x)$ satisfies the following two-scale equation,

$$\phi_j(x) = \phi_{j+1}(x) + \sum_{l=0}^{K_j-1} \phi_j((2l+1)h_{j+1})\phi_{j+1}(x - (2l+1)h_{j+1}).$$

Proof.   From (4.1.8)

$$\phi_j(kh_j) = \frac{1}{K_j} \sum_{l=0}^{K_j-1} \frac{Z_l^j(kh_j)}{Z_l^j(0)}$$

$$= \frac{1}{K_j} \sum_{l=0}^{K_j-1} e^{-ilkh_j}$$

$$= \delta_{k,0}$$

hence (1) is true.

Since $\{\phi_j(\cdot - kh_j)\}_{k=0}^{K_j-1}$ is a linearly independent system, $\phi_j(\cdot - kh_j) \in V_j$, from (4.1.2), $\dim V_j = K_j$, hence we have (2).

(3) follows from $V_j \subset V_{j+1}$ and the cardinal interpolatory property of $\phi_j(x)$.

$\square$

§4.2   The Periodic Cardinal Interpolatory Wavelets (PCIW)

We consider a function $R_l^j(x)$:

$$R_l^j(x) = \{C_l^{j+1}\tilde{Z}_l^{j+1}(x) - C_{K_j+l}^{j+1}\tilde{Z}_{K_j+l}^{j+1}(x)\}e^{ilh_{j+1}} \tag{4.2.1}$$

$l = 0, 1, \cdots, K_j - 1$, where

$$C_l^{j+1} = \frac{\|Z_{K_j+l}^{j+1}\|}{\|Z_l^j\|}.$$

Lemma 4.2.1 R_l^j is orthogonal to Z_λ^j, i.e.,

$$\langle R_l^j(\cdot), Z_\lambda^j(\cdot)\rangle = 0, \quad 0 \le l, \lambda \le K_j - 1. \tag{4.2.2}$$

Proof.   From (4.1.4), we have

$$Z_\lambda^j(x) = K_j \sum_{n\in\mathbf{Z}} d_{nK_j-\lambda} e^{i(nK_j-\lambda)x}$$

$$= K_j \left\{ \sum_{n\in\mathbf{Z}} d_{2nK_j-\lambda} e^{i(2nK_j-\lambda)x} + \sum_{n\in\mathbf{Z}} d_{(2n-1)K_j-\lambda} e^{i((2n-1)K_j-\lambda)x} \right\}$$

$$= \tfrac{1}{2}\{Z_\lambda^{j+1}(x) + Z_{K_j+\lambda}^{j+1}(x)\}. \tag{4.2.3}$$

On the other hand

$$\langle R_l^j(\cdot), Z_\lambda^j(\cdot)\rangle = \tfrac{1}{2}\langle [C_l^{j+1}\tilde{Z}_l^{j+1}(\cdot) - C_{K_j+l}^{j+1}\tilde{Z}_{K_j+l}^{j+1}(\cdot)]e^{ilh_{j+1}},$$

$$Z_\lambda^{j+1}(\cdot) + Z_{K_j+\lambda}^{j+1}(\cdot)\rangle$$

$$= \tfrac{1}{2}e^{ilh_{j+1}}[C_l^{j+1}\|Z_\lambda^{j+1}\| - C_{K_j+l}^{j+1}\|Z_{K_j+l}^{j+1}\|]\delta_{\lambda,l}$$

$$= 0.$$

$$\square$$

Since

$$R_l^j(h_{j+1}) = \{C_l^{j+1}\tilde{Z}_l^{j+1}(h_{j+1}) - C_{K_j+l}^{j+1}\tilde{Z}_{K_j+l}^{j+1}(h_{j+1})\}e^{ilh_{j+1}}$$

$$= C_l^{j+1}\tilde{Z}_l^{j+1}(0) + C_{K_j+l}^{j+1}\tilde{Z}_{K_j+l}^{j+1}(0)$$

$$> 0,$$

we may define the periodic cardinal interpolatory wavelets $L_j(x)$ as follows:

$$L_j(x) = \frac{1}{K_j} \sum_{l=0}^{K_j-1} \frac{R_l^j(x)}{R_l^j(h_{j+1})} \tag{4.2.4}$$

Definition 4.2.1

$$W_j = \text{span}\{R_k^j(\cdot) | k = 0, 1, \cdots, K_j - 1\}. \tag{4.2.5}$$

we have the following:

Theorem 4.2.1
(1) $\{L_j(\cdot - lh_j)\}_{l=0}^{K_j-1}$ is a basis for W_j;
(2) $W_j \perp V_j, V_{j+1} = V_j \oplus W_j$;
(3) $L_j(kh_j + h_{j+1}) = \delta_{k,0}, k = 0, 1, \cdots, K_j - 1$;
(4) $L_j(x)$ satisfies the following equation

$$L_j(x) = \phi_{j+1}(x - h_{j+1}) + \sum_{l=0}^{K_j-1} L_j(lh_j)\phi_{j+1}(x - lh_j).$$

Proof.   From (4.2.4) and (4.2.5), we know that

$$L_j(\cdot - kh_j) \in W_j, \quad \{L_j(\cdot - lh_j)\}_{l=0}^{K_j-1}$$

is linearly independent. Since

$$L_j(kh_j + h_{j+1}) = \frac{1}{K_j} \sum_{l=0}^{K_j-1} \frac{R_l^j(kh_j + h_{j+1})}{R_l^j(h_{j+1})}$$

$$= \frac{1}{K_j} \sum_{l=0}^{K_j-1} e^{-iklh_j}$$

$$= \delta_{0,k},$$

it follows that (3) and (1) are true. (2) follows from (4.2.2); and

$$\dim V_j = \dim W_j = K_j.$$

It is easy to obtain (4) by a simple calculation.   □

Since $\phi_j(x)$ and $L_j(x)$ are the linear combination of $g(x)$ and its translations, the regularity of $\phi_j(x)$ and $L_j(x)$ are the same as that of $g(x)$. Therefore we can construct periodic interpolatory wavelets with any order of regularity.

§4.3   Symmetry of Scaling Functions and Wavelets

Assume that $g(x)$ is real-valued and symmetric about the origin, that is

$$g(x) = g(-x) = \overline{g(x)}. \tag{4.3.1}$$

This implies

$$d_n = d_{-n} = \overline{d_n}, \qquad n \in \mathbf{Z}, \tag{4.3.2}$$

where $\{d_n\}_n$ are the Fourier coefficients of $g(x)$.

From (4.3.1) and (4.1.3), it is easy to establish the following equalities:

$$\overline{Z_l^j}(x) = Z_{-l}^j(x) = Z_{K_j-l}^j(x) = Z_l^j(-x) \tag{4.3.3}$$

and

$$C_l^{j+1} = C_{-l}^{j+1} = C_{K_{j+1}-l}^{j+1}. \tag{4.3.4}$$

Theorem 4.3.1   Let ϕ_j and L_j be the functions defined in (4.1.8) and (4.2.4) respectively, and g satisfies (4.3.1). Then

(i) $\phi_j(x)$ is real-valued and $\phi_j(-x) = \phi_j(x)$;

(ii) $L_j(x)$ is real-valued and $L_j(h_{j+1} + x) = L_j(h_{j+1} - x)$.

Proof.   From (4.1.8), (4.3.3) and $Z_{K_j}^j(x) = Z_0^j(x)$, we have

$$\overline{\phi_j(x)} = \phi_j(x)$$

and

$$\phi_j(-x) = \frac{1}{K_j} \sum_{l=0}^{K_j-1} \frac{\overline{Z_l^j(-x)}}{\overline{Z_l^j(0)}}$$

$$= \frac{1}{K_j} \sum_{l=0}^{K_j-1} \frac{\overline{Z_l^j(x)}}{\overline{Z_l^j(0)}}$$

$$= \overline{\phi_j(x)}$$

$$= \phi_j(x),$$

thus (i) is true.

To prove (ii), we note that

$$\overline{R_l^j}(x) = \{C_l^{j+1}\overline{\tilde{Z}_l^{j+1}}(x) - C_{K_j+l}^{j+1}\overline{\tilde{Z}_{K_j+l}^{j+1}}(x)\}e^{-ilh_{j+1}}$$

$$= \{C_l^{j+1}\tilde{Z}_{-l}^{j+1}(x) - C_{K_j+l}^{j+1}\tilde{Z}_{-K_j-l}^{j+1}(x)\}e^{-ilh_{j+1}}$$

$$= \{C_{-l}^{j+1}\tilde{Z}_{-l}^{j+1}(x) - C_{K_j-l}^{j+1}\tilde{Z}_{K_j-l}^{j+1}(x)\}e^{-ilh_{j+1}}$$

$$= \{-C_{K_{j+1}-l}^{j+1}\tilde{Z}_{K_{j+1}-l}^{j+1}(x) + C_{K_j-l}^{j+1}\tilde{Z}_{K_j-l}^{j+1}(x)\}e^{i(K_j-l)h_{j+1}}$$

$$= R_{K_j-l}^j(x). \tag{4.3.5}$$

By (4.3.5) and $R_{K_j}^j(x) = R_0^j(x)$, we have

$$\overline{L_j(x)} = \frac{1}{K_j}\sum_{l=0}^{K_j-1}\frac{\overline{R_l^j(x)}}{R_l^j(h_{j+1})}$$

$$= \frac{1}{K_j}\sum_{l=0}^{K_j-1}\frac{R_{K_j-l}^j(x)}{R_{K_j-l}^j(h_{j+1})}$$

$$= \frac{1}{K_j}\sum_{l=1}^{K_j}\frac{R_l^j(x)}{R_l^j(h_{j+1})}$$

$$= \frac{1}{K_j}\sum_{l=0}^{K_j-1}\frac{R_l^j(x)}{R_l^j(h_{j+1})}$$

$$= L_j(x). \tag{4.3.6}$$

Using (4.3.6) and $R_{-l}^j(h_{j+1}+x) = R_l^j(h_{j+1}-x)$, it follows

$$L_j(h_{j+1}-x) = \frac{1}{K_j}\sum_{l=0}^{K_j-1}\frac{R_l^j(h_{j+1}-x)}{R_l^j(h_{j+1})}$$

$$= \frac{1}{K_j}\sum_{l=0}^{K_j-1}\frac{R_{-l}^j(h_{j+1}+x)}{R_{-l}^j(h_{j+1})}$$

$$= \overline{L_j(h_{j+1}+x)}$$

$$= L_j(h_{j+1}+x).$$

Thus $L_j(x)$ is symmetric about the point h_{j+1}. $\qquad\square$

§4.4   Dual Scaling Functions and Dual Wavelets

In this section we construct the Dual scaling functions and dual wavelets.

The following lemma is important for construction of the Dual scaling functions.

Lemma 4.4.1   Let ϕ_j be the function defined in (4.1.8). Define a polynomial

$$P_j(z) := a_0 + a_1 z + \cdots + a_{K_j-1} z^{K_j-1} \qquad (4.4.1)$$

where $a_\nu = \langle \phi_j(\cdot), \phi_j(\cdot - \nu h_j) \rangle, \nu = 0, \cdots, K_j - 1$, and two matrices

$$G_j(z) := (b_{l,k})_{l,k=0}^{K_j-1}; \qquad (4.4.2)$$

$$F := \frac{1}{\sqrt{K_j}} (\omega^{lk})_{l,k=0}^{K_j-1}, \qquad (4.4.3)$$

where

$$\omega = e^{ih_j}, \quad b_{l,k} = \langle \phi_j(\cdot - lh_j), \phi_j(\cdot - kh_j) \rangle.$$

Then G_j is invertible and

$$G_j^{-1} = F \wedge_j^{-1} F^* \qquad (4.4.4)$$

where

$$\Lambda_j = \mathrm{diag}\{P_j(1), P_j(\omega), \cdots, P_j(\omega^{K_j-1})\}$$

and the star * denotes the complex conjugate.

Proof.   Since $b_{l,k} = a_{k-l}$, this implies that G_j is circulant. So G_j can be diagonalized by F, i.e.

$$G_j = F\Lambda_j F^*$$

(refer to [Ga]).

We now prove $P_j(\omega^r) \neq 0$ for $r = 0, \cdots, K_j - 1$.

By (4.1.8) and the orthogonality of $\{Z_l^j\}$, we have

$$a_k = \frac{1}{K_j^2} \sum_{l,n=0}^{K_j-1} \left\langle \frac{Z_l^j(\cdot)}{Z_l^j(0)}, \frac{Z_n^j(\cdot - kh_j)}{Z_n^j(0)} \right\rangle$$

$$= \frac{1}{K_j^2} \sum_{l,n=0}^{K_j-1} \left\langle \frac{Z_l^j(\cdot)}{Z_l^j(0)}, \frac{Z_n^j(\cdot)}{Z_n^j(0)} \right\rangle e^{-iknh_j}$$

$$= \frac{1}{K_j^2} \sum_{l=0}^{K_j-1} \frac{\|Z_l^j\|^2}{|Z_l^j(0)|} e^{-iklh_j},$$

hence, for $r = 0, 1, \cdots, K_j - 1$,

$$P_j(\omega^r) = \sum_{k=0}^{K_j-1} a_k \omega^{rk}$$

$$= \frac{1}{K_j^2} \sum_{l=0}^{K_j-1} \frac{\|Z_l^j\|^2}{|Z_l^j(0)|^2} \sum_{k=0}^{K_j-1} e^{i(r-l)kh_j}$$

$$= \frac{1}{K_j} \frac{\|Z_r^j\|^2}{|Z_r^j(0)|^2}$$

$$> 0.$$

Therefore, G_j is invertible, and

$$G_j^{-1} = F\Lambda_j^{-1}F^*.$$

                                                           □

Theorem 4.4.1   If

$$\tilde{\phi}_j(x) := eF\Lambda_j^{-1}F^* \begin{pmatrix} \phi_j(x) \\ \phi_j(x - h_j) \\ \vdots \\ \phi_j(x - (K_j-1)h_j) \end{pmatrix}, \qquad (4.4.5)$$

where $e = (1, 0, \cdots, 0) \in R^{K_j}$, and Λ_j and F are defined as in the lemma above, then
$\{\tilde{\phi}_j(\cdot - kh_j)\}_{k=0}^{K_j-1}$ is a dual basis for $\{\phi_j(\cdot - kh_j)\}_{k=0}^{K_j-1}$, i.e.,

$$\langle \tilde{\phi}_j(\cdot - kh_j), \phi_j(\cdot - lh_j) \rangle = \delta_{k,l} \qquad (4.4.6)$$

for $k, l = 0, \cdots, K_j - 1$.

Proof. Let

$$\pi := \mathrm{Circ}(0, 1, 0, \cdots, 0) = \begin{pmatrix} 0 & 1 & 0 & \cdots & 0 \\ 0 & 0 & 1 & \cdots & 0 \\ 0 & 0 & 0 & \cdots & 0 \\ \vdots & \vdots & \vdots & \ddots & \vdots \\ 1 & 0 & 0 & \cdots & 0 \end{pmatrix}.$$

Since $F\Lambda_j^{-1}F^*$ is a circulant matrix,

$$F\Lambda_j^{-1}F^*\pi = \pi F\Lambda_j^{-1}F^*.$$

Hence

$$\tilde{\phi}_j(x - kh_j) = eF\Lambda_j^{-1}F^*(\phi_j(x - kh_j), \phi_j(x - (k+1)h_j), \cdots,$$

$$\phi_j(x - (k + K_j - 1)h_j))^T$$

$$= eF\Lambda_j^{-1}F^* \cdot \pi^k \cdot (\phi_j(x), \phi_j(x - h_j), \cdots, \phi_j(x - (K_j - 1)h_j))^T$$

$$= e\pi^k F\Lambda_j^{-1}F^* \cdot (\phi_j(x), \phi_j(x - h_j), \cdots, \phi_j(x - (K_j - 1)h_j))^T.$$

Therefore

$$\langle \tilde{\phi}_j(\cdot - kh_j), \phi_j(\cdot - lh_j) \rangle$$

$$= e\pi^k F\Lambda_j^{-1}F^* \begin{pmatrix} \langle \phi_j(\cdot), \phi_j(\cdot - lh_j) \rangle \\ \langle \phi_j(\cdot - h_j), \phi_j(\cdot - lh_j) \rangle \\ \vdots \\ \langle \phi_j(\cdot - (K_j - 1)h_j, \phi_j(\cdot - lh_j) \rangle \end{pmatrix}$$

or, equivalently

$$(\langle \tilde{\phi}_j(\cdot - kh_j), \phi_j(\cdot - lh_j) \rangle)_{k,l=0}^{K_j-1}$$

$$= I \cdot F\Lambda_j^{-1}F^* \cdot G_j$$

$$= I.$$

we have thus proved that (4.4.6) is true.   □

The dual wavelet $\tilde{L}_j$ also can be defined.

Define

$$\tilde{L}_j(x) = e \cdot F \cdot \mathrm{diag}\{Q_j(1)^{-1}, (Q_j(\omega))^{-1}, \cdots, (Q_j(\omega^{K_j-1}))^{-1}\}$$

$$\cdot \begin{pmatrix} L_j(x) \\ L_j(x - h_j) \\ \vdots \\ L_j(x - (K_j - 1)h_j) \end{pmatrix}, \tag{4.4.7}$$

where

$$Q_j(z) = \sum_{k=0}^{K_j-1} \langle L_j(\cdot), L_j(\cdot - kh_j) \rangle z^k.$$

Theorem 4.4.2

$$\langle \tilde{L}_j(\cdot - kh_j), L_j(\cdot - lh_j) \rangle = \delta_{k,l}, \quad k, l = 0, 1, \cdots, K_j - 1. \tag{4.4.8}$$

Proof.   Since

$$Q_j(\omega^r) = \frac{K_j[(C_r^{j+1})^2 + (C_{K_j+r}^{j+1})^2]}{|R_r^2(h_{j+1})|^2}$$

$$> 0,$$

the proof of Theorem 4.4.2 is analogous to that of Theorem 4.4.1.   □

In contrast to Theorem 4.3.1, we have

Theorem 4.4.3   Assume that $g(x)$ is real-valued and symmetric about the origin. Then $\tilde{\phi}_j$ and $\tilde{L}_j$ are symmetric about the origin.

Proof.   For simplicity, we write $\tilde{\phi}_j(x)$ as

$$\tilde{\phi}_j(x) = \sum_{k=0}^{K_j-1} C_k \phi_j(x - kh_j).$$

By using the symmetry and periodicity of ϕ_j, we have

$$\tilde{\phi}_j(-x) = \sum_{k=0}^{K_j-1} C_k \phi_j(-x - kh_j)$$

$$= C_0 \phi_j(x) + \sum_{k=1}^{K_j-1} C_k \phi_j(x + kh_j)$$

$$= C_0 \phi_j(x) + \sum_{k=1}^{K_j-1} C_k \phi_j(x - (K_j - k)h_j)$$

$$= C_0 \phi_j(x) + \sum_{k=1}^{K_j-1} C_{K_j-k} \phi_j(x - kh_j).$$

Since

$$\text{Circ}(C_0, C_1, \cdots, C_{K_j-1}) = F\Lambda^{-1}F^*$$

and

$$\Lambda = \text{diag}\{P_j(1), P_j(\omega), \cdots, P_j(\omega^{K_j-1})\}$$

is real, $\text{Circ}(C_0, C_1, \cdots, C_{K_j-1})$ is Hermitian and $C_{K_j-k} = C_k$. Therefore

$$\tilde{\phi}_j(-x) = \sum_{k=0}^{K_j-1} C_k \phi_j(x - kh_j) = \tilde{\phi}_j(x).$$

An analogous discussion leads to $\tilde{L}_j(x) = \tilde{L}_j(-x)$. $\square$

§4.5   Algorithms

We shall deduce the decomposition and reconstruction formulas corresponding to this kind of periodic wavelets.

Let $f \in V_{j+1}$. Since $V_{j+1} = V_j \oplus W_j$, we can expand the function f in two kinds of expressions

$$f(x) = \sum_{l=0}^{K_{j+1}-1} t_l^{j+1} \phi_{j+1}(x - lh_{j+1}) \tag{4.5.1}$$

and

$$f(x) = \sum_{l=0}^{K_j-1} t_l^j \phi_j(x - lh_j) + \sum_{l=0}^{K_j-1} S_l^j L_j(x - lh_j) \qquad (4.5.2)$$

where

$$t_l^{j+1} = f(lh_{j+1}). \qquad (4.5.3)$$

Since the two functions $\phi_{j+1}(x)$ and $\phi_{j+1}(x - h_{j+1})$ belong to V_{j+1}, we have the expansions

$$\phi_{j+1}(x) = \sum_{l=0}^{K_j-1} \{a_l^j \phi_j(x - lh_j) + b_l^j L_j(x - lh_j)\}, \qquad (4.5.4)$$

$$\phi_{j+1}(x - h_{j+1}) = \sum_{l=0}^{K_j-1} \{p_l^j \phi_j(x - lh_j) + q_l^j L_j(x - lh_j)\}. \qquad (4.5.5)$$

From (4.5.4) and (4.5.5) we have

$$\phi_{j+1}(x - 2kh_{j+1}) = \sum_{l=0}^{K_j-1} \{a_{l-k}^j \phi_j(x - lh_j) + b_{l-k}^j L_j(x - lh_j)\}, \qquad (4.5.6)$$

$$\phi_{j+1}(x - (2k+1)h_{j+1}) = \sum_{l=0}^{K_j-1} \{p_{l-k}^j \phi_j(x - lh_j) + q_{l-k}^j L_j(x - lh_j)\}. \qquad (4.5.7)$$

Using (4.5.1), (4.5.6) and (4.5.7)

$$f(x) = \sum_{k=0}^{K_j-1} t_{2k}^{j+1} \phi_{j+1}(x - 2kh_{j+1}) + \sum_{k=0}^{K_j-1} t_{2k+1}^{j+1} \phi_{j+1}(x - (2k+1)h_{j+1})$$

$$= \sum_{k=0}^{K_j-1} t_{2k}^{j+1} \sum_{l=0}^{K_j-1} \{a_{l-k}^j \phi_j(x - lh_j) + b_{l-k}^j L_j(x - lh_j)\}$$

$$+ \sum_{k=0}^{K_j-1} t_{2k+1}^{j+1} \sum_{l=0}^{K_j-1} \{p_{l-k}^j \phi_j(x - lh_j) + q_{l-k}^j L_j(x - lh_j)\}$$

$$= \sum_{l=0}^{K_j-1} \left\{ \sum_{k=0}^{K_j-1} [a_{l-k}^j t_{2k}^{j+1} + p_{l-k}^j t_{2k+1}^{j+1}] \right\} \phi_j(x - lh_j)$$

$$+ \sum_{l=0}^{K_j-1} \left\{ \sum_{k=0}^{K_j-1} [b_{l-k}^j t_{2k}^{j+1} + q_{l-k}^j t_{2k+1}^{j+1}] \right\} L_j(x - lh_j).$$

$$(4.5.8)$$

Comparing (4.5.8) with (4.5.2) we have:

Decomposition Formulas

$$t_l^j = \sum_{k=0}^{K_j-1} [a_{l-k}^j t_{2k}^{j+1} + p_{l-k}^j t_{2k+1}^{j+1}], \qquad (4.5.9)$$

$$S_l^j = \sum_{k=0}^{K_j-1} [b_{l-k}^j t_{2k}^{j+1} + q_{l-k}^j t_{2k+1}^{j+1}]. \qquad (4.5.10)$$

Since ϕ_j, L_j belong to V_{j+1}, by Theorem 4.1.1 (1), we have

$$\phi_j(x) = \sum_{l=0}^{K_{j+1}-1} \phi_j(lh_{j+1})\phi_{j+1}(x - lh_{j+1}), \qquad (4.5.11)$$

$$L_j(x) = \sum_{l=0}^{K_{j+1}-1} L_j(lh_{j+1})\phi_{j+1}(x - lh_{j+1}). \qquad (4.5.12)$$

From (4.5.2), (4.5.11) and (4.5.12)

$$f(x) = \sum_{l=0}^{K_{j+1}} \left\{ \sum_{k=0}^{K_j-1} t_k^j \phi_j((l - 2k)h_{j+1}) + \sum_{k=0}^{K_j-1} S_k^j L_j((l - 2k)h_{j+1}) \right\}$$
$$\times \phi_{j+1}(x - lh_{j+1}), \qquad (4.5.13)$$

and by comparing the coefficients in (4.5.13) with those in (4.5.1), we obtain:

Reconstruction Formulas

$$t_{2l}^{j+1} = t_l^j + \sum_{k=0}^{K_j-1} S_k^j L_j((l - k)h_j),$$
$$\qquad (4.5.14)$$
$$t_{2l+1}^{j+1} = S_l^j + \sum_{k=0}^{K_j-1} t_k^j \phi_j((l - k)h_j + h_{j+1}).$$

We are now going to compute a_l^j, b_l^j, p_l^j and q_l^j. From (4.5.4) and the dual property (4.4.6) it follows that

$$a_l^j = \langle \tilde{\phi}_j(\cdot - lh_j), \phi_{j+1}(\cdot) \rangle. \qquad (4.5.15)$$

Recall that

$$\tilde{\phi}_j(x) = \sum_{k=0}^{K_{j+1}-1} \tilde{\phi}_j(kh_{j+1})\phi_{j+1}(x - kh_{j+1}) \in V_{j+1}.$$

Substituting $\tilde{\phi}_j(x)$ into (4.5.15), we obtain the formula for computing a_l^j. b_l^j, p_l^j and q_l^j can be obtained similarly. Thus we have:

Formulas for computing a_l^j, b_l^j, p_l^j and q_l^j

$$a_l^j = \sum_{k=0}^{K_{j+1}-1} \tilde{\phi}_j((k - 2l)h_{j+1})\langle \phi_{j+1}(\cdot - kh_{j+1}), \phi_{j+1}(\cdot)\rangle, \qquad (4.5.16)$$

$$b_l^j = \sum_{k=0}^{K_{j+1}-1} \tilde{L}_j((k - 2l)h_{j+1})\langle \phi_{j+1}(\cdot - kh_{j+1}), \phi_{j+1}(\cdot)\rangle, \qquad (4.5.17)$$

$$p_l^j = \sum_{k=0}^{K_{j+1}-1} \tilde{\phi}_j((k - 2l + 1)h_{j+1})\langle \phi_{j+1}(\cdot - kh_{j+1}), \phi_{j+1}(\cdot)\rangle, \qquad (4.5.18)$$

$$q_l^j = \sum_{k=0}^{K_{j+1}-1} \tilde{L}_j((k - 2l + 1)h_{j+1})\langle \phi_{j+1}(\cdot - kh_{j+1}), \phi_{j+1}(\cdot)\rangle. \qquad (4.5.19)$$

In the following we present two different approaches to investigating the localization property of periodic interpolatory scaling functions. The first is the spline approach; it shows that interpolatory functions decay exponentially in one period. The method we introduce here can be applied to more general classes of functions (see [CC]). We shall demonstrate this method for Bernoulli splines. The second approach is angular localization. This method was initiated by Narcowich and Ward [NW].

§4.6   Localization of PISF via Spline Approach

In Chapter II we defined a class of functions $S_n(h_j)$ such that if $g(x) \in S_n(h_j)$ then $g(x)$ is a spline of degree n with knots $\{lh_j | l \in \mathbb{Z}\}$ if n is odd, or $\left\{\left(l + \frac{1}{2}\right) h_j \middle| l \in \mathbb{Z}\right\}$ if n is even. We also defined

$$\overset{\circ}{S}_n(h_j) = \{g | g(\cdot) \in S_n(h_j), g(\cdot) \text{ is of period } T\} \qquad (4.6.1)$$

Let $y = (y_\nu)$ be a prescribed sequence of numbers, where the subscript ν ranges over all integers. The problem of finding a function $F(x)$ which belongs to some prescribed linear space S and satisfies the relation

$$F(\nu) = y_\nu \tag{4.6.2}$$

for all $\nu \in Z$, is called the cardinal interpolation problem and is denoted by the symbol $CIP(y, S)$.

For $s \geq 0$ consider the class

$$F_s = \{F(x)|F(x) \in C, F(x) = 0(|x|^s) \quad \text{as} \quad x \to \pm\infty\}. \tag{4.6.3}$$

In particular, F_0 is the class of bounded and continuous functions. We may also describe the class $F^* = \bigcup_{s>0} F_s$ as the class of functions of power growth.

The following lemma shows that the interpolation condition (4.6.2) sets up a one-to-one correspondence between the elements of the two classes

$$S_n(1) \cap F^* \quad \text{and} \quad Y^*, \tag{4.6.4}$$

where $Y^* = \bigcup_{s \geq 0} Y_s$ and

$$Y_s = \{y = (y_\nu)| \ y_\nu = O(|\nu|^s), \nu \to \pm\infty\}. \tag{4.6.5}$$

Lemma 4.6.1   The

$$CIP(g, S_n(1) \cap F_s) \tag{4.6.6}$$

has a unique solution if and only if $y \in Y_s$.

Corollary 4.6.1   Given the unit sequence $\delta = (\delta_\nu)$, where $\delta_\nu = 1$ if $\nu = 0$ and 0 otherwise, then the $CIP(\delta, S_n(1) \cap F_0)$ has a unique solution, which we denote by $F_n(x)$ (the fundamental spline function), such that

$$F_n(\nu) = \delta_\nu, \quad \text{for all} \quad \nu \in Z. \tag{4.6.7}$$

Lemma 4.6.2   The function $F_n(x)$ satisfies the following inequality

$$|F_n(x)| < C_n e^{-\gamma_n |x|}, \quad \text{all} \quad x \in R. \tag{4.6.8}$$

(see *p*.211 Note 2).

Define

$$\gamma_{l,j}(x) = \sum_{\gamma \in Z} F_n(h_j^{-1} x - l - \nu K_j). \tag{4.6.9}$$

We have the following:

Theorem 4.6.1   For each $l, 0 \leq l \leq K_j - 1$, the function $\gamma_{l,j}(x)$ enjoys the following properties:

(i) $\gamma_{l,j}(\lambda h_j) = \delta_{l,\lambda}, \quad 0 \leq l, \lambda \leq K_j - 1$,

(ii) $\Gamma = \{\gamma_{l,j}\}_{l=0}^{K_j-1}$ is a basis of $\overset{\circ}{S}_n(h_j)$,

(iii) $\gamma_{K_j-1,j}(x)$ is even and is symmetric about the point $M := T/2$,

(iv) there exist two constants C_1 and C_2 such that

$$|\gamma_{K_j-1,j}(x)| \leq C_1 e^{-2^j C_2 |x-M|}, \quad \forall x \in [0,T]$$

(see *p*.211, Note 3).

Proof.   (i)

$$\begin{aligned}
\gamma_{l,j}(\lambda h_j) &= \sum_{\nu \in Z} F_n(h_j^{-1} h_j \lambda - l - \nu K_j) \\
&= \sum_{\nu \in Z} \delta_{\lambda-l,\nu K_j} \\
&= \delta_{\lambda-l,0} \\
&= \delta_{\lambda,l}.
\end{aligned}$$

(ii) Since $F_n(x) \in S_n(1)$ we have $F_n(x - \nu) \in S_n(1)$, where ν is any integer, $F_n(h_j^{-1} x - l - \nu K_j) \in S_n(h_j)$, and

$$\gamma_{l,j} = \sum_{\nu \in Z} F_n(h_j^{-1} x - l - \nu K_j) \in \overset{\circ}{S}_n(h_j).$$

If there are constants $\{C_l\}$ such that

$$\sum_{l=0}^{K_j-1} C_l \gamma_{l,j}(x) = 0, \quad \forall x \in [0,T]$$

then

$$\sum_{l=0}^{K_j-1} C_l \gamma_{l,j}(\lambda h_j) = C_\lambda = 0, \quad \lambda = 0, \cdots, K_j - 1.$$

$\{\gamma_{l,j}\}_{l=0}^{K_j-1}$ is a class of linearly independent functions. Thus Γ is a basis of $\overset{\circ}{S}_n(h_j)$.

(iii) From the definition of $\gamma_{K_{j-1},j}(x)$, we have:

$$\gamma_{K_{j-1},j}(-x) = \sum_{\nu \in Z} F_n(-h_j^{-1}x - K_{j-1} - \nu K_j)$$

$$= \sum_{\nu \in Z} F_n(h_j^{-1}x + K_{j-1} + \nu K_j)$$

since $F_n(x)$ is even [Sch2, p.414, (3,4)]. Set $\nu = -\lambda - 1$, we have:

$$\gamma_{K_{j-1},j}(-x) = \sum_{\lambda \in Z} F_n(h_j^{-1}x + K_{j-1} - (\lambda + 1)K_j)$$

$$= \sum_{\lambda \in Z} F_n(h_j^{-1}x - K_{j-1} - \lambda K_j) = \gamma_{K_{j-1},j}(x).$$

So $\gamma_{K_{j-1},j}(x)$ is even.

$$\gamma_{K_{j-1},j}(M - a) = \sum_{\lambda \in Z} F_n(h_j^{-1}(M - a) - K_{j-1} - \nu K_j)$$

$$= \sum_{\lambda \in Z} F_n(h_j^{-1}(a - M) + K_{j-1} + \nu K_j)$$

$$= \sum_{\nu \in Z} F_n(h_j^{-1}(a + M) - 2Mh_j^{-1} + K_{j-1} + \nu K_j)$$

$$= \sum_{\lambda \in Z} F_n(h_j^{-1}(a + M) - K_{j-1} - \lambda K_j)$$

$$= \gamma_{K_{j-1},j}(M + a),$$

hence $\gamma_{K_{j-1},j}(x)$ is symmetric with respect to the point M.

iv) From (iii), we only need to consider the case $x \in [M, T]$. From (4.6.8)

$$|\gamma_{K_{j-1},j}(x)| = \left| \sum_{\nu \in Z} F_n(h_j^{-1}x - K_{j-1} - \nu K_j) \right|$$

$$\leq \sum_{\nu \in Z} C_n e^{-(2^j \frac{K}{T} r_n |x - K_{j-1}h_j - \nu T|)} \qquad (4.6.10)$$

$$= I_1 + I_2 + I_3.$$

We now estimate the three quantities.

$$I_1 = \sum_{\nu \geq 1} C_n e^{-[2^j \frac{K}{T} \gamma_n (\nu T - (x - M))]}$$

$$\leq \sum_{\nu \geq 1} C_n e^{-2^j \frac{K}{T} \gamma_n [(\nu - 1)T + (x - M)]}$$

$$= C_n e^{-2^j \frac{K}{T} \gamma_n (x - M)} \cdot \frac{e^{2^j K}}{e^{2^j K} - 1}$$

$$= C_j^1 \xi$$

where

$$C_j^1 = \frac{e^{2^j K}}{[e^{2^j K} - 1]}, \quad \xi = C_n e^{-2^j \frac{K}{T} \gamma_n (x - M)},$$

C_j^1 is bounded for $j \geq 0$. Since

$$I_2 = \xi, \quad I_3 = I_1,$$

we have

$$I_1 + I_2 + I_3 = (2C_j^1 + 1)C_n e^{-2^j \frac{K}{T} \gamma_n (x - M)}$$

$$\leq C_1 e^{-2^j C_2 |x - M|}$$

where C_1, C_2 are constants. $\qquad\qquad\square$

Remark C_1, C_2 are independent of j. From (iv) $\gamma_{K_{j-1}, j}(x)$ decays very fast for each $x \in [0, T]$ as $j \to \infty$.

In the previous paragraph we have presented the properties of the periodic cardinal interpolatory spline function (abbreviated as PCIS) $\gamma_{K_{j-1}, j}(x)$. Its formal presentation comes from $F_n(x)$ via formula (4.6.9); Schoenberg's result [Sch2] yields the construction of $F_n(x)$ theoretically.

Here we construct the PCIS through the Bernoulli polynomials, which is different from (4.6.9) and is easier to manipulate.

The Bernoulli polynomial ϕ_m is a polynomial of degree m defined inductively on m over the interval $I = [0, T]$:

$$\phi_0(x) = 1, \quad \phi_m'(x) = \phi_{m-1}(x), \quad \int_0^T \phi_m(t)dt = 0 \quad (m \geq 1). \quad (4.6.11)$$

Thus we have

$$\phi_1(x) = x - \frac{T}{2}$$

$$\phi_2(x) = \frac{1}{2}(x^2 - Tx) + \frac{T^2}{12}$$

$$\phi_3(x) = \frac{x^3}{6} - \frac{T}{4}x^2 + \frac{T^2}{12}x \qquad (4.6.12)$$

$$\phi_4(x) = \frac{x^4}{24} - \frac{T}{12}x^3 + \frac{T^2}{24}x^2 - \frac{T^4}{720}$$

(see Note 4).

Periodizing ϕ_m with period T, we have the Fourier expansion for the extended ϕ_m

$$\phi_m(x) = -2\left(\frac{T}{2\pi}\right)^m \sum_{\nu \geq 1} \frac{1}{\nu^m} \cos\left(\frac{2\nu\pi}{T}x - \frac{m\pi}{2}\right) \qquad (4.6.13)$$

$m = 1, 2, \cdots$.

Lemma 4.6.3   The function $\phi_m(x - \nu h_j)$ defined in (4.6.13) has the following properties:

(i) $\phi_m(x - \nu h_j) \in C^{m-2}(R)$,

(ii) $\phi_m(x - \nu h_j)$ has knots $\{\nu h_j + \lambda T\}_{\lambda \in Z}$,

(iii) $\phi_m(x - \nu h_j)$ is polynomial of degree m on interval $(\nu h_j + \lambda T, \nu h_j + (\lambda + 1)T)$, for all $\lambda \in Z$.

(iv) When m is even,

$$\{1, \phi_m(\cdot - \nu h_j) - \phi_m(\cdot - (K_j - 1)h_j), \ \nu = 0, \cdots, K_j - 2\}$$

is a basis of $\overset{\circ}{S}_m(h_j)$ (see (4.6.1)). When m odd, the basis is

$$\{1, \phi_m(\cdot - \nu h_j - \frac{1}{2}h_j) - \phi_m(\cdot - (K_j - 1)h_j - \frac{1}{2}h_j), \ \nu = 0, \cdots, K_j - 2\}.$$

Proof.   The periodic extension of $\phi_1(x)$ (see (4.6.12)) is of $C^{-1}(R)$, by (4.6.11) $\phi'_m(x) = \phi_{m-1}(x)(m \geq 2)$, we have (i).

Since the periodic extension of $\phi_m(x)$ has knots $\{\lambda T\}_{\lambda \in Z}$, therefore, we have (ii). (iii) is evident.

(iv) is a result proved by G. Meinardus. (see Note 5).   □

Corollary 4.6.2   A function f belongs to $\overset{\circ}{S}_m(h_j)$ if and only if it can be written in the form

$$f(x) = a_0 + \sum_{\nu=0}^{K_j-1} C_\nu \phi_m \left(x - \nu h_j - \frac{\varepsilon}{2} h_j \right) \qquad (4.6.14)$$

where $C_0 + \cdots + C_{K_j-1} = 0, \varepsilon = 0$ if m even and 1 if m odd.   (4.6.15)

Moreover,

$$a_0 = \frac{1}{T} \int_0^T f(t)dt, \quad C_\nu = -\frac{1}{T}[f^{(m-1)}(\nu h_j + 0) - f^{(m-1)}(\nu h_j - 0)], \quad (4.6.16)$$

$\nu = 0, \cdots, K_j - 1$.

The above corollary tells us that we can use the Bernoulli spline function to construct any function in $\overset{\circ}{S}_m(h_j)$, but the inverse is not true, which means that they belong to different kinds of spline spaces.

Now we start to construct the PCIS.

Define

$$E_l^{m,j}(x) := \sum_{\lambda=0}^{K_j-1} \phi_m \left(x - \lambda h_j - \frac{\varepsilon}{2} h_j \right) e^{il\lambda h_j \frac{2\pi}{T}}. \qquad (4.6.17)$$

We write $\phi_m(x)$ as

$$\phi_m(x) = -\left(\frac{T}{2\pi} \right)^m \cdot \sum_{\nu \neq 0} \frac{1}{(i\nu)^m} e^{i\frac{2\nu\pi}{T}x}, \qquad (4.6.18)$$

and substitute (4.6.18) into (4.6.17) and then evaluate at $x = rh_j + \frac{\varepsilon}{2}h_j$. Then

$$E_l^{m,j} \left(rh_j + \frac{\varepsilon}{2}h_j \right) = e^{ilrh_j \frac{2\pi}{T}} E_0^{m,j} \qquad (4.6.19)$$

where

$$E_0^{m,j} = -\left(\frac{T}{2\pi} \right)^m K_j \sum_{\nu \neq 0} \frac{1}{[i(l + \nu K_j)]^m}, \quad 0 \leq l \leq K_j - 1. \qquad (4.6.20)$$

It is clear that

$$E_l^{m,j}(0) \neq 0, \quad m \text{ even, for } \quad l : 0 \leq l \leq K_j - 1 \qquad (4.6.21)$$

$$E_l^{m,j}(0) \neq 0, \quad m \text{ odd, for } \quad l : 1 \leq l \leq K_j - 1. \qquad (4.6.22)$$

Set

$$b_l^{m,j}(x) := \frac{E_l^{m,j}(x)}{E_l^{m,j}(0)}, \quad 0 < l \leq K_j - 1. \qquad (4.6.23)$$

This leads to

$$b_l^{m,j}(0) = 1 \quad \text{for} \quad 0 < l \leq K_j - 1. \qquad (4.6.24)$$

The following theorem gives the structure of PCIS.

Theorem 4.6.2   Let $b_l^{m,j}(x)$ be the functions in (4.6.23). Then

$$\rho^{m,j}(x) := \frac{1}{K_j}\left(1 + \sum_{l=1}^{K_j-1} b_l^{m,j}(x)\right), \quad x \in R \qquad (4.6.25)$$

is the unique periodic Lagrange function in $\overset{\circ}{S}_m(h_j)$ which satisfies

$$\rho^{m,j}\left(\nu h_j + \frac{\varepsilon}{2}h_j\right) = \delta_{0,\nu}, \quad 0 \leq \nu \leq K_j - 1 \qquad (4.6.26)$$

where $\varepsilon = 0$ if m even, 1 if m odd.

Proof.   Taking $x = \nu h_j + \frac{\varepsilon}{2}h_j$ in (4.6.25) yields

$$\rho^{m,j}\left(\nu h_j + \frac{\varepsilon}{2}h_j\right) = \frac{1}{K_j}\left(1 + \sum_{l=1}^{K_j-1} b_l^{m,j}\left(\nu h_j + \frac{\varepsilon}{2}h_j\right)\right)$$

$$= \frac{1}{K_j}\left(1 + \sum_{l=1}^{K_j-1} e^{il\nu h_j \frac{2\pi}{T}}\right)$$

$$= \delta_{\nu,0}.$$

This is (4.6.26).

Now we prove $\rho^{m,j}$ belongs to $\overset{\circ}{S}_m(h_j)$.

$$
\begin{aligned}
\rho^{m,j}(x) &= \frac{1}{K_j}\left(1 + \sum_{l=1}^{K_j-1} b_l^{m,j}(x)\right) \\
&= \frac{1}{K_j}\left(1 + \sum_{l=1}^{K_j-1} E_l^{m,j}(x)/E_l^{m,j}(0)\right) \\
&= \frac{1}{K_j}\left\{1 + \sum_{l=1}^{K_j-1} \frac{1}{E_l^{m,j}(0)} \sum_{\lambda=0}^{K_j-1} \phi_m\left(x - \lambda h_j - \frac{\varepsilon}{2}h_j\right)e^{il\lambda h_j \frac{2\pi}{T}}\right\} \\
&= \frac{1}{K_j}\left\{1 + \sum_{\lambda=0}^{K_j-1} d_\lambda \phi_m\left(x - \lambda h_j - \frac{\varepsilon}{2}h_j\right)\right\}
\end{aligned}
$$

$$(4.6.27)$$

where

$$
\begin{aligned}
d_\lambda &= \sum_{l=1}^{K_j-1} \frac{1}{E_l^{m,j}(0)} e^{il\lambda h_j \frac{2\pi}{T}}, \\
\sum_{\lambda=0}^{K_j-1} d_\lambda &= \sum_{l=1}^{K_j-1} \frac{1}{E_l^{m,j}(0)} \sum_{\lambda=0}^{K_j-1} e^{il\lambda h_j \frac{2\pi}{T}} \\
&= \sum_{l=1}^{K_j-1} \frac{1}{E_l^{m,j}(0)} K_j \delta_{l,0} \\
&= 0.
\end{aligned}
$$

$$(4.6.28)$$

From Corollary 4.6.1, (4.6.27) and (4.6.28), we assert that $\rho^{m,j}(x)$ belongs to $\overset{\circ}{S}_m(h_j)$. $\square$

Remark   The function $\gamma_{l,j}(x)$ in (4.6.9) can be constructed as follows

$$
\gamma_{l,j}(x) := \rho^{n,j}\left(x - lh_j + \frac{\varepsilon}{2}h_j\right), \qquad 0 \le l \le K_j - 1. \qquad (4.6.29)
$$

§4.7   Localization of PISF via Circular Variance Approach

We now investigate the localization of the periodic interpolation scaling function (PISF) constructed in §4.1 from a different point of view. The method which we shall use was introduced by Narcowich and Ward [NW].

For a 2π-periodic continuous differentiable function f whose L^2 norm is 1, set

$$\tau(f) := \int_0^{2\pi} e^{it}|f(t)|^2 dt. \tag{4.7.1}$$

If we regard $|f(t)|^2$ as a distribution of mass on the unit circle, then $\tau(f)$ is the center of mass of the distribution. The corresponding variance is

$$\int_0^{2\pi} |e^{it} - \tau(f)|^2 |f(t)|^2 dt = 1 - |\tau(f)|^2. \tag{4.7.2}$$

The size of $1 - |\tau(f)|^2$ is a good measure of how well localized $|f(t)|^2$ is about $\tau(f)$. For example, if $|f(t)|^2$ goes to a point mass located at t_0, then $\tau(f)$ approaches e^{it_0}, and $1 - |\tau(f)|^2$ approaches 0. Conversely, if $1 - |\tau(f)|^2 = 0$, then the distribution corresponds to a point mass located at $\tau(f)$.

Through the above discussion we now define the circular variance of a continuous periodic function f as follows:

$$\mathrm{Var}(f) := 1 - |\tau(f)|. \tag{4.7.3}$$

Let

$$f(x) = \sum_{k \in Z} d_k e^{ikx}, \quad d_k > 0, \quad \{d_k\} \in l^1.$$

Using $f(x)$ we can construct the PISF φ_j (see (4.1.8)). What we should do is to explore how good the function φ_j is.

It is clear that φ_j's norm is not 1. So we should change the definition of $\tau(f)$ to

$$\tau(f) = \frac{1}{2\pi} \int_0^{2\pi} \frac{e^{it}|f(t)|^2}{\|f\|^2} \, dt. \tag{4.7.4}$$

Denote

$$\tilde{S}_l^j = K_j^{-1} Z_l^j(0) = \sum_{n \in Z} d_{nK_j - l}. \tag{4.7.5}$$

Then from (4.1.8) and (4.7.5), we have

$$\frac{1}{2\pi} \int_0^{2\pi} e^{it}|\phi_j(t)|^2 dt = \left(\frac{1}{K_j}\right)^2 \sum_{l=-K_{j-1}}^{K_{j-1}-1} \sum_{n \in Z} \frac{d_{nK_{j-1}-l}}{\tilde{S}_l^j} \frac{d_{nK_{j-1}-l-1}}{\tilde{S}_{l+1}^j}, \tag{4.7.6}$$

and

$$K_j^2 \|\phi_j\|^2 = \sum_{l=-K_{j-1}}^{K_{j-1}-1} \sum_{n\in\mathbf{Z}} \left(\frac{d_{nK_j-l}}{\tilde{S}_l^j}\right)^2. \tag{4.7.7}$$

From (4.7.6) and (4.7.7)

$$
\begin{aligned}
\mathrm{Var}(\phi_j) &= \frac{1}{K_j^2\|\phi_j\|^2} \sum_{l=K_{j-1}}^{K_{j-1}-1} \sum_{n\in\mathbf{Z}} \left\{ \left(\frac{d_{nK_j-l}}{\tilde{S}_l^j}\right)^2 - \frac{d_{nK_j-l}}{\tilde{S}_l^j}\frac{d_{nK_j-l-1}}{\tilde{S}_{l+1}^j} \right\} \\
&= \frac{1}{K_j^2\|\phi_j\|^2} \sum_{l=K_{j-1}}^{K_{j-1}-1} \sum_{n\in\mathbf{Z}} \left(\frac{d_{nK_j-l}}{\tilde{S}_l^j}\right)\left(\frac{d_{nK_j-l}}{\tilde{S}_l^j} - \frac{d_{nK_j-l-1}}{\tilde{S}_{l+1}^j}\right) \\
&\le \frac{1}{K_j\|\phi_j\|} \left\{ \sum_{l=-K_{j-1}}^{K_{j-1}-1} \sum_{n\in\mathbf{Z}} \left|\frac{d_{nK_j-l-1}}{\tilde{S}_{l+1}^j} - \frac{d_{nK_j-l}}{\tilde{S}_l^j}\right|^2 \right\}.
\end{aligned}
$$
$$\tag{4.7.8}$$

Now we establish the following

Lemma 4.7.1   Suppose $\left\{\frac{d_{k+1}}{d_k} - 1\right\} \in l^2$. Then there exists a constant M_0 independent of j, such that

$$\sum_{l=-K_{j-1}}^{K_{j-1}-1} \sum_{n\in\mathbf{Z}} \left|\frac{d_{nK_j-l-1}}{\tilde{S}_{l+1}^j} - \frac{d_{nK_j-l}}{\tilde{S}_l^j}\right|^2 \le M_0. \tag{4.7.9}$$

Proof.

$$
\begin{aligned}
&\sum_{l=-K_{j-1}}^{K_{j-1}-1} \sum_{n\in\mathbf{Z}} \left|\frac{d_{nK_j-l-1}}{\tilde{S}_{l+1}^j} - \frac{d_{nK_j-l}}{\tilde{S}_l^j}\right|^2 \\
&= \sum_{l=K_{j-1}}^{K_{j-1}-1} \sum_{n\in\mathbf{Z}} \left|\frac{d_{nK_j-l-1}}{\tilde{S}_{l+1}^j} - \frac{d_{nK_j-l-1}}{\tilde{S}_l^j} + \frac{d_{nK_j-l-1}}{\tilde{S}_l^j} - \frac{d_{nK_j-l}}{\tilde{S}_l^j}\right|^2 \\
&\le 2\sum_{l=-K_{j-1}}^{K_{j-1}-1} \sum_{n\in\mathbf{Z}} \left(\left|\frac{d_{nK_j-l-1}}{\tilde{S}_{l+1}^j} - \frac{d_{nK_j-l-1}}{\tilde{S}_l^j}\right|^2 + \left|\frac{d_{nK_j-l-1}}{\tilde{S}_l^j} - \frac{d_{nK_j-l}}{\tilde{S}_l^j}\right|^2 \right) \\
&\le 2\sum_{l=-K_{j-1}}^{K_{j-1}-1} \sum_{n\in\mathbf{Z}} \left[d_{nK_j-l-1}^2 \left(\frac{1}{\tilde{S}_l^j} - \frac{1}{\tilde{S}_{l+1}^j}\right)^2 + \left(\frac{d_{nK_j-l-1}}{d_{nK_j-l}} - 1\right)^2 \right]
\end{aligned}
$$

$$\leq 2M_1 + 2 \sum_{l=-K_{j-1}}^{K_{j-1}-1} \sum_{n \in Z} d_{nK_j-l-1}^2 \left(\frac{1}{\tilde{S}_l^j} - \frac{1}{\tilde{S}_{l+1}^j} \right)^2$$

$$\leq 2M_1 + 2 \sum_{l=-K_{j-1}}^{K_{j-1}-1} \sum_{n \in Z} \left(\frac{d_{nK_j-l-1}}{\tilde{S}_l^j \tilde{S}_{l+1}^j} \right)^2 \left(\tilde{S}_{l+1}^j - \tilde{S}_l^j \right)^2$$

$$\leq 2M_1 + 2 \sum_{l=-K_{j-1}}^{K_{j-1}-1} \left(\frac{1}{\tilde{S}_l^j} \right)^2 \left[\sum_{m \in Z} (d_{mK_j-l-1} - d_{mK_j-l}) \right]^2$$

$$= 2M_1 + 2 \sum_{l=-K_{j-1}}^{K_{j-1}-1} \left(\frac{1}{\tilde{S}_l^j} \right)^2 \sum_{m \in Z} \left(\frac{d_{mK_j-l-1}}{d_{mK_j-l}} - 1 \right)^2 \cdot \sum_{m \in Z} d_{mK_j-l}^2$$

$$\leq M_0$$

□

From (4.7.8) and (4.7.9), it follows

$$\text{Var}(\phi_j) \leq \frac{M_0}{(K_j \|\phi_j\|)}. \qquad (4.7.10)$$

Theorem 4.7.1   Suppose that

$$\left\{ \frac{d_{k+1}}{d_k} - 1 \right\} \in l^2.$$

Then we have

$$\lim_{j \to \infty} \text{Var}(\varphi_j) = 0. \qquad (4.7.11)$$

Proof.   From the definition of $\phi_j(t)$, we have

$$K_j^2 \|\phi_j\|^2 = \sum_{l=-K_{j-1}}^{K_{j-1}-1} \sum_{n \in Z} \frac{d_{nK_j-l}^2}{(\tilde{S}_l^j)^2}. \qquad (4.7.12)$$

Fixing l, we shall show

$$\lim_{j \to \infty} \frac{d_{-l}}{\tilde{S}_l^j} = 1. \qquad (4.7.13)$$

In fact,

$$d_{nK_j-l} = \frac{1}{2\pi} \int_0^{2\pi} f(t)e^{ilt} \cdot e^{-inK_jt}dt$$

$$= \frac{1}{2\pi K_j} \int_0^{2\pi K_j} f\left(\frac{\tau}{K_j}\right) e^{il\frac{\tau}{K_j}} e^{-in\tau}d\tau$$

$$= \frac{1}{2\pi} \int_0^{2\pi} \frac{1}{K_j} \left[\sum_{\nu=0}^{K_j-1} f\left(\frac{2\pi\nu+y}{K_j}\right) e^{il\frac{2\pi\nu+y}{K_j}} \right] e^{-iny}dy.$$

The Fourier coefficients of the function

$$p_l(y) = \frac{1}{K_j} \sum_{\nu=0}^{K_j-1} f\left(\frac{2\pi\nu+y}{K_j}\right) e^{il\frac{2\pi\nu+y}{K_j}}$$

are $\{d_{nK_j-l}\}_{n\in\mathbf{Z}}$. Therefore

$$p_l(y) = \sum_{n\in\mathbf{Z}} d_{nK_j-l}e^{iny},$$

$$\sum_{n\in\mathbf{Z}} d_{nK_j-l} = p_l(0) = \frac{1}{K_j} \sum_{\nu=0}^{K_j-1} f\left(\frac{2\pi\nu}{K_j}\right) e^{il\frac{2\pi\nu}{K_j}}. \qquad (4.7.14)$$

From (4.7.14), it is clear that

$$\tilde{S}_l^j = \sum_n d_{nK_j-l} \to \frac{1}{2\pi} \int_0^{2\pi} f(t)e^{ilt}dt = d_{-l}$$

as $j \to \infty$, (4.7.13) is obtained.

We now prove that for any given $N > 0$ there exists j_0 such that

$$K_j^2\|\phi_j\|^2 > N \quad \text{for all} \quad j > j_0. \qquad (4.7.15)$$

In fact, for any given $N > 0$, there is $j_1(N)$ such that

$$\frac{d_{-l}}{\tilde{S}_l^j} > \frac{1}{\sqrt{2}}, \quad \text{for} \quad |l| \le N \qquad (4.7.16)$$

if $j > j_1(N)$.

Let j_2, j_0 be two integers such that $K_{j_2} - 1 \geq N$ and $j_0 = \max\{j_1, j_2\}$. Then from (4.7.12)

$$K_j^2 \|\phi_j\|^2 \geq \sum_{l=-K_{j-1}}^{K_{j-1}-1} \frac{d_{-l}^2}{(\tilde{S}_l^j)^2}$$

$$> \sum_{|l| \leq N} \frac{d_{-l}^2}{(\tilde{S}_l^j)^2}$$

$$> \tfrac{1}{2}(2N + 1)$$

$$> N$$

for all $j > j_0$. We thus have (4.7.15).

From (4.7.10), (4.7.15) it follows (4.7.11) since N is chosen arbitrarily.

□

Corollary 4.7.1   Suppose the sequence $\{d_k\}$ satisfies the following conditions:

$$\inf_{\substack{|l| \leq K_{j-1} \\ j \geq 0}} \frac{d_{-l}}{\tilde{S}_l^j} \geq C > 0 \qquad (4.7.17)$$

and

$$\left\{\frac{d_{k+1}}{d_k} - 1\right\} \in l^2. \qquad (4.7.18)$$

Then

$$\mathrm{Var}(\phi_j) = O\left(\frac{1}{\sqrt{K_j}}\right) \qquad (j \to \infty). \qquad (4.7.19)$$

Proof.   From (4.7.12) and (4.7.17)

$$K_j^2 \|\phi_j\|^2 \geq CK_j. \qquad (4.7.20)$$

Combining (4.7.20) with (4.7.10), we obtain (4.7.19).   □

§4.8   Local Properties of PCIW.

In this section we study the local properties of PCIW. We prove that the PCIW has very good localization when the level j gets larger.

Throughout this section, C will denote a constant.

We shall establish the following main result.

Theorem 4.8.1   Assume that $\{d_k\} \in l^1$, (4.7.17) and (4.7.18) are valid. Then

$$|\text{Var}(L_j)| = O\left(\frac{1}{\sqrt{K_j}}\right) \quad \text{as} \quad j \to \infty. \tag{4.8.1}$$

For proving (4.8.1) we need some lemmas.

Lemma 4.8.1   Let $L_j(x)$ be the function in (4.2.4). Then

$$L_j(x) = \frac{1}{K_j} \sum_{l=-K_{j-1}}^{K_{j-1}-1} \left[\frac{M_l^{j+1} Z_l^{j+1}(x) - (M_l^{j+1})^{-1} Z_{K_j+l}^{j+1}(x)}{M_l^{j+1} Z_l^{j+1}(0) + (M_l^{j+1})^{-1} Z_{K_j+l}^{j+1}(0)} \right] e^{ilh_{j+1}} \tag{4.8.2}$$

where

$$M_l^{j+1} = \frac{\|Z_{K_j+l}^{j+1}\|}{\|Z_l^{j+1}\|}. \tag{4.8.3}$$

Proof. Since $K_{j+1} = 2K_j$ and $Z_{K_{j+1}+l}^{j+1}(x) = Z_l^{j+1}(x)$, from (4.2.1) we have

$$R_l^j(x) = \frac{1}{\|Z_l^j\|}\{M_l^{j+1} Z_l^{j+1}(x) - (M_l^{j+1})^{-1} Z_{K_j+l}^{j+1}(x)\} e^{ilh_{j+1}} \tag{4.8.4}$$

and

$$R_l^j(h_{j+1}) = \left\{ C_l^{j+1} \frac{Z_l^{j+1}(h_{j+1})}{\|Z_l^{j+1}\|} - C_{K_j+l}^{j+1} \frac{Z_{K_j+l}^{j+1}(h_{j+1})}{\|Z_{K_j+l}^{j+1}\|} \right\} e^{ilh_{j+1}}$$

$$= C_l^{j+1} \frac{Z_l^{j+1}(0)}{\|Z_l^{j+1}\|} + C_{K_j+l}^{j+1} \frac{Z_{K_j+l}^{j+1}(0)}{\|Z_{K_j+l}^{j+1}\|} \tag{4.8.5}$$

$$= \frac{1}{\|Z_l^j\|}\{M_l^{j+1} Z_l^{j+1}(0) + (M_l^{j+1})^{-1} Z_{K_j+l}^{j+1}(0)\}.$$

By using (4.8.4), (4.8.5), (4.2.4) and $M_{l+K_j}^{j+1} = (M_l^{j+1})^{-1}$, we can easily obtain (4.8.2). $\square$

Lemma 4.8.2   The integral averages of functions $e^{ix}|L_j(x)|^2$ and $|L_j(x)|^2$ are given by formulas (a) and (b) respectively:

(a) $\frac{1}{2\pi} \int_0^{2\pi} e^{ix}|L_j(x)|^2 dx = \frac{e^{ih_{j+1}}}{K_j^2} \left[\sum_{l=-K_{j-1}}^{K_{j-1}-1} \sum_{n\in\mathbb{Z}} X_{n,l} + \sum_{n\in\mathbb{Z}}(Y_n + Z_n) \right]$

where

$$X_{n,l} = \frac{M_l^{j+1} M_{l+1}^{j+1} d_{nK_{j+1}-l} d_{nK_{j+1}-(l+1)}}{G_l^{j+1} \cdot G_{l+1}^{j+1}}$$

$$+ \frac{(M_l^{j+1} M_{l+1}^{j+1})^{-1} d_{nK_{j+1}-(K_j+l)} d_{nK_{j+1}-(K_j+l+1)}}{G_l^{j+1} \cdot G_{l+1}^{j+1}}$$

$$Y_n = \frac{d_{nK_{j+1}+K_{j-1}} d_{(n+1)K_{j+1}-(K_j+K_{j-1}-1)} M_{-K_{j-1}}^{j+1} (M_{K_{j-1}-1}^{j+1})^{-1}}{G_{-K_{j-1}}^{j+1} \cdot G_{K_{j-1}-1}^{j+1}}$$

$$Z_n = \frac{d_{nK_{j+1}-K_{j-1}+1} d_{nK_{j+1}-K_{j-1}} M_{K_{j-1}-1}^{j+1} (M_{-K_{j-1}}^{j+1})^{-1}}{G_{-K_{j-1}}^{j+1} \cdot G_{K_{j-1}-1}^{j+1}}$$

and

$$G_\nu^{j+1} = M_\nu^{j+1} \tilde{S}_\nu^{j+1} + (M_\nu^{j+1})^{-1} \tilde{S}_{K_j+\nu}^{j+1}.$$

(b)

$$\|L_j\|^2 = \frac{1}{2\pi} \int_0^{2\pi} |L_j(x)|^2 dx$$

$$= \frac{1}{K_j^2} \sum_{l=-K_{j-1}}^{K_{j-1}-1} \sum_{n\in\mathbb{Z}} \frac{(M_l^{j+1})^2 d_{nK_{j+1}-l}^2 + (M_l^{j+1})^{-2} d_{nK_{j+1}-K_j-l}^2}{(G_l^{j+1})^2}.$$

Proof.   (a) From (4.8.2), it is easily to show

$$\frac{1}{2\pi} \int_0^{2\pi} e^{ix}|L_j(x)|^2 dx = A_1 + A_2 + A_3 + A_4 \tag{4.8.6}$$

where

$$A_1 = \frac{1}{K_j^2} \sum_{l,k=-K_{j-1}}^{K_{j-1}-1} \frac{1}{2\pi} \int_0^{2\pi} e^{ix} M_l^{j+1} M_k^{j+1} Z_l^{j+1}(x) \overline{Z_k^{j+1}}(x) dx \cdot P_{l,k}^{j+1}$$

$$A_2 = \frac{1}{K_j^2} \sum_{l,k=-K_{j-1}}^{K_{j-1}-1} \frac{1}{2\pi} \int_0^{2\pi} e^{ix} M_l^{j+1} (M_k^{j+1})^{-1} Z_l^{j+1}(x) \overline{Z_{K_j+k}^{j+1}(x)} dx \cdot P_{l,k}^{j+1}$$

$$A_3 = \frac{1}{K_j^2} \sum_{l,k=-K_{j-1}}^{K_{j-1}-1} \frac{1}{2\pi} \int_0^{2\pi} e^{ix} (M_l^{j+1})^{-1} M_k^{j+1} Z_{K_j+l}^{j+1}(x) \overline{Z_k^{j+1}}(x) dx \cdot P_{l,k}^{j+1}$$

$$A_4 = \frac{1}{K_j^2} \sum_{l,k=-K_{j-1}}^{K_{j-1}-1} \frac{1}{2\pi} \int_0^{2\pi} e^{ix} (M_l^{j+1} M_k^{j+1})^{-1} Z_{K_j+l}^{j+1}(x) \overline{Z_{K_j+k}^{j+1}}(x) dx \cdot P_{l,k}^{j+1}$$

and

$$P_{l,k}^{j+1} = \frac{e^{i(l-k)h_{j+1}}}{H_l^{j+1} \cdot H_k^{j+1}}, \qquad H_\nu^{j+1} = M_\nu^{j+1} Z_\nu^{j+1}(0) + (M_\nu^{j+1})^{-1} Z_{K_j+\nu}^{j+1}(0).$$

Now we comput A_1. From (4.1.4), we have

$$\frac{1}{2\pi} \int_0^{2\pi} Z_l^{j+1}(x) \overline{Z_k^{j+1}}(x) e^{ix} dx = K_{j+1}^2 \sum_{n\in Z} d_{nK_{j+1}-(k+1)} d_{nK_{j+1}-k}.$$

Consequently,

$$A_1 = \frac{1}{K_j^2} \sum_{k=-K_{j-1}}^{K_{j-1}-1} \frac{M_k^{j+1} M_{k+1}^{j+1} \sum_{n\in Z} d_{nK_{j+1}-k} d_{nK_{j+1}-(k+1)} e^{ih_{j+1}}}{G_k^{j+1} \cdot G_{k+1}^{j+1}}.$$

$$(4.8.7)$$

Similarly, we have

$$A_2 = \frac{e^{ih_{j+1}}}{K_j^2} \sum_{n\in Z} \frac{d_{nK_{j+1}+K_{j-1}} d_{(n+1)K_{j+1}-(K_j+K_{j-1}-1)} M_{-K_{j-1}}^{j+1} (M_{K_{j-1}-1}^{j+1})^{-1}}{G_{K_{j-1}-1}^{j+1} \cdot G_{-K_{j-1}}^{j+1}}$$

$$(4.8.8)$$

$$A_3 = \frac{e^{ih_{j+1}}}{K_j^2} \sum_{n\in Z} \frac{d_{nK_{j+1}-K_{j-1}+1} d_{nK_{j+1}-K_{j-1}} M_{K_{j-1}-1}^{j+1} (M_{-K_{j-1}}^{j+1})^{-1}}{G_{-K_{j-1}}^{j+1} \cdot G_{K_{j-1}-1}^{j+1}}$$

$$(4.8.9)$$

and

$$A_4 = \frac{e^{ih_{j+1}}}{K_j^2} \sum_{l=-K_{j-1}}^{K_{j-1}-1} \sum_{n \in \mathbb{Z}} \frac{d_{nK_{j+1}-(K_j+l)} d_{nK_{j+1}-(K_j+l+1)} (M_l^{j+1} M_{l+1}^{j+1})^{-1}}{G_l^{j+1} \cdot G_{l+1}^{j+1}}$$

$$(4.8.10)$$

From (4.8.7)–(4.8.10) and (4.8.6), (a) follows.

(b) The computation of $\|L_j\|^2$, is analogous to that in (a).   □

Lemma 4.8.3   Suppose $\left\{ \dfrac{d_{n+1}}{d_n} - 1 \right\} \in l^2$ and $\{d_n\} \in l^2$. Then there are two constants C_1 and C_2 independent of j such that

$$\sum_{l=-K_{j-1}}^{K_{j-1}-1} \left| \frac{(M_l^{j+1})^2}{(M_{l+1}^{j+1})^2} - 1 \right|^2 \leq C_1$$

$$(4.8.11)$$

and

$$\sum_{l=-K_{j-1}}^{K_{j-1}-1} \left| \frac{(M_{l+1}^{j+1})^2}{(M_l^{j+1})^2} - 1 \right|^2 \leq C_2.$$

$$(4.8.12)$$

Proof.   From (4.8.3) and the inequality $|a + b|^2 \leq 2(|a|^2 + |b|^2)$, we have

$$\sum_{l=-K_{j-1}}^{K_{j-1}-1} \left| \frac{(M_l^{j+1})^2}{(M_{l+1}^{j+1})^2} - 1 \right|$$

$$= \sum_{l=-K_{j-1}}^{K_{j-1}-1} \left| \frac{Q_{K_j+l}}{Q_l} \cdot \frac{Q_{l+1}}{Q_{K_j+l+1}} - 1 \right|^2$$

$$= \sum_{l=-K_{j-1}}^{K_{j-1}-1} \left| \left(\frac{Q_{l+1}}{Q_l} - 1 \right) \left(\frac{Q_{K_j+l}}{Q_{K_j+l+1}} \right) + \frac{Q_{K_j+l}}{Q_{K_j+l+1}} - 1 \right|^2$$

$$\leq 2 \sum_{l=-K_{j-1}}^{K_{j-1}-1} \left\{ \left| \left(\frac{Q_{l+1}}{Q_l} - 1 \right) \left(\frac{Q_{K_j+l}}{Q_{K_j+l+1}} \right) \right|^2 + \left| \frac{Q_{K_j+l}}{Q_{K_j+l+1}} - 1 \right|^2 \right\}$$

$$= I_1 + I_2$$

$$(4.8.13)$$

where $Q_\nu := \sum_{n \in \mathbb{Z}} d_{nK_{j+1}-\nu}^2.$

Now we estimate I_1 and I_2.

$$I_1 = 2 \sum_{l=-K_{j-1}}^{K_{j-1}-1} \left| \left(\frac{Q_{l+1}}{Q_l} - 1 \right) \frac{Q_{K_j+l}}{Q_{K_j+l+1}} \right|^2$$

$$= 2 \sum_{l=-K_{j-1}}^{K_{j-1}-1} \left| \left(\frac{Q_{l+1}}{Q_l} - 1 \right) \left(\frac{Q_{K_j+l}}{Q_{K_j+l+1}} - 1 \right) + \left(\frac{Q_{l+1}}{Q_l} - 1 \right) \right|^2.$$

Using the inequality $|a+b|^2 \leq 2(|a|^2 + |b|^2)$, then

$$I_1 \leq 4 \sum_{l=-K_{j-1}}^{K_{j-1}-1} \left\{ \left| \left(\frac{Q_{l+1}}{Q_l} - 1 \right) \left(\frac{Q_{K_j+l}}{Q_{K_j+l+1}} - 1 \right) \right|^2 + \left| \frac{Q_{l+1}}{Q_l} - 1 \right|^2 \right\}$$

$$\leq 4 \left[\sum_{l=-K_{j-1}}^{K_{j-1}-1} \left| \frac{Q_{l+1}}{Q_l} - 1 \right|^2 \right] \left[\sum_{l=-K_{j-1}}^{K_{j-1}-1} \left| \frac{Q_{K_j+l}}{Q_{K_j+l+1}} - 1 \right|^2 \right]$$

$$+ 4 \sum_{l=-K_{j-1}}^{K_{j-1}-1} \left| \frac{Q_{l+1}}{Q_l} - 1 \right|^2, \tag{4.8.14}$$

Now turn to estimate I_2.

$$I_2 = 2 \sum_{l=-K_{j-1}}^{K_{j-1}-1} \left| \frac{Q_{K_j+l}}{Q_{K_j+l+1}} - 1 \right|^2$$

$$= 2 \sum_{l=-K_{j-1}}^{K_{j-1}-1} \left| \frac{Q_{K_j+l} - Q_{K_j+l+1}}{Q_{K_j+l+1}} \right|^2.$$

By the definition of Q_ν above, then

$$I_2 = 2 \sum_{l=-K_{j-1}}^{K_{j-1}-1} \left| \frac{\sum_{n\in Z} d_{nK_{j+1}-K_j-l-1}^2 \left(\frac{d_{nK_{j+1}-K_j-l}^2}{d_{nK_{j+1}-K_j-l-1}^2} - 1 \right)}{Q_{K_j+l+1}} \right|^2.$$

From the Schwartz inequality

$$I_2 \le 2 \sum_{l=-K_{j-1}}^{K_{j-1}-1} \left| \frac{\left[\sum_{n\in Z} d^4_{nK_{j+1}-K_j-l-1}\right]^{\frac{1}{2}} \left[\sum_{n\in Z} \left(\frac{d^2_{nK_{j+1}-K_j-l}}{d^2_{nK_{j+1}-K_j-l-1}} - 1\right)^2\right]^{\frac{1}{2}}}{Q_{K_j+l+1}} \right|^2$$

$$= 2 \sum_{l=-K_{j-1}}^{K_{j-1}-1} \frac{\sum_{n\in Z} d^4_{nK_{j+1}-K_j-l-1}}{[\sum_{n\in Z} d^2_{nK_{j+1}-K_j-l-1}]^2} \cdot \sum_{n\in Z} \left[\frac{d^2_{nK_{j+1}-K_j-l}}{d^2_{nK_{j+1}-K_j-l-1}} - 1\right]^2$$

$$\le 2 \sum_{l=-K_{j-1}}^{K_{j-1}-1} \sum_{n\in Z} \left[\frac{d_{nK_{j+1}-K_j-l}}{d_{nK_{j+1}-K_j-l-1}} - 1\right]^2 \left[\frac{d_{nK_{j+1}-K_j-l}}{d_{nK_{j+1}-K_j-l-1}} + 1\right]^2$$

$$\le C. \tag{4.8.15}$$

Since

$$\left\{\frac{d_{n+1}}{d_n} - 1\right\} \in l^2$$

by the assumption, the last factor

$$\left[\frac{d_{nK_{j+1}-K_j-l}}{d_{nK_{j+1}-K_j-l-1}} + 1\right]$$

is bounded.

Now return to (4.8.14). By using the method in the estimation for I_1, the boundedness of I_1 is clear. Thus, by (4.8.13), we have the estimation in (4.8.11).

Now we prove (4.8.12).

Let λ be the number:

$$\lambda := \sum_{l=-K_{j-1}}^{K_{j-1}-1} \left| \frac{(M^{j+1}_{l+1})^2}{(M^{j+1}_l)^2} - 1 \right|^2$$

$$= \sum_{l=-K_{j-1}}^{K_{j-1}-1} \left| \frac{Q_{K_j+l+1} \cdot Q_l}{Q_{l+1} \cdot Q_{K_j+l}} - 1 \right|^2$$

$$= \sum_{l=-K_{j-1}}^{K_{j-1}-1} \left| \left[\frac{Q_l}{Q_{l+1}} - 1\right] \frac{Q_{K_j+l+1}}{Q_{K_j+l}} + \frac{Q_{K_j+l+1}}{Q_{K_j+l}} - 1 \right|^2.$$

Using the inequality $|a + b|^2 \leq 2(|a|^2 + |b|^2)$, then

$$\lambda \leq 2 \sum_{l=-K_{j-1}}^{K_{j-1}-1} \left\{ \left| \left[\frac{Q_l}{Q_{l+1}} - 1 \right] \frac{Q_{K_j+l+1}}{Q_{K_j+l}} \right|^2 + \left| \frac{Q_{K_j+l+1}}{Q_{K_j+l}} - 1 \right|^2 \right\}$$

$$= 2 \sum_{l=-K_{j-1}}^{K_{j-1}-1} \left\{ \left[\left(\frac{Q_l}{Q_{l+1}} - 1 \right) \times \left(\frac{Q_{K_j+l+1}}{Q_{K_j+l}} - 1 \right) + \left(\frac{Q_l}{Q_{l+1}} - 1 \right) \right]^2 \right\}$$

$$+ 2 \sum_{l=-K_{j-1}}^{K_{j-1}-1} \left| \frac{Q_{K_j+l+1}}{Q_{K_j+l}} - 1 \right|^2$$

Using the same inequality above again,

$$\lambda \leq 4 \sum_{l=-K_{j-1}}^{K_{j-1}-1} \left\{ \left(\frac{Q_l}{Q_{l+1}} - 1 \right)^2 \left(\frac{Q_{K_j+l+1}}{Q_{K_j+l}} - 1 \right)^2 + \left(\frac{Q_l}{Q_{l+1}} - 1 \right)^2 \right\}$$

$$+ 2 \sum_{l=-K_{j-1}}^{K_{j-1}-1} \left| \frac{Q_{K_j+l+1}}{Q_{K_j+l}} - 1 \right|^2$$

By Schwartz inequality,

$$\lambda \leq 4 \left\{ \sum_{l=K_{j-1}}^{K_{j-1}-1} \left[\frac{Q_l}{Q_{l+1}} - 1 \right]^4 \right\}^{\frac{1}{2}} \left\{ \sum_{l=K_{j-1}}^{K_{j-1}-1} \left[\frac{Q_{K_j+l+1}}{Q_{K_j+l}} - 1 \right]^4 \right\}^{\frac{1}{2}}$$

$$+ 4 \sum_{l=-K_{j-1}}^{K_{j-1}-1} \left\{ \left| \frac{Q_l}{Q_{l+1}} - 1 \right|^2 + \frac{1}{2} \left| \frac{Q_{K_j+l+1}}{Q_{K_j+l}} - 1 \right|^2 \right\}$$

$$\leq 4 \left\{ \sum_{l=K_{j-1}}^{K_{j-1}-1} \left| \frac{Q_l}{Q_{l+1}} - 1 \right|^2 \right\} \left\{ \sum_{l=K_{j-1}}^{K_{j-1}-1} \left| \frac{Q_{K_j+l+1}}{Q_{K_j+l}} - 1 \right|^2 \right\}$$

$$+ 4 \sum_{l=-K_{j-1}}^{K_{j-1}-1} \left\{ \left| \frac{Q_l}{Q_{l+1}} - 1 \right|^2 + \frac{1}{2} \left| \frac{Q_{K_j+l+1}}{Q_{K_j+l}} - 1 \right|^2 \right\} \leq C,$$

here the last inequality follows from an estimate similar to that for I_2. $\square$

Lemma 4.8.4   Assume

$$\left\{ \frac{d_n}{d_{n+1}} - 1 \right\} \in l^2.$$

Then there exists a constant C independent of j such that

$$\sum_{l=-K_{j-1}}^{K_{j-1}-1} \left| \frac{\tilde{S}_{K_j l+1}^{j+1}}{\tilde{S}_{K_j+l}^{j}} - 1 \right|^2 \leq C. \tag{4.8.16}$$

Proof.   Denote

$$I := \sum_{l=-K_{j-1}}^{K_{j-1}-1} \left| \frac{\tilde{S}_{K_j+l+1}^{j+1}}{\tilde{S}_{K_j+l}^{j+1}} - 1 \right|^2,$$

By the definition of $\tilde{S}_l^j$ (see (4.7.5)),

$$I = \sum_{l=-K_{j-1}}^{K_{j-1}-1} \left| \frac{\sum_{n\in\mathbb{Z}}(d_{nK_{j+1}-K_j-l-1} - d_{nK_{j+1}-K_j-l})}{\sum_{n\in\mathbb{Z}} d_{nK_{j+1}-K_j-l}} \right|^2$$

$$= \sum_{l=-K_{j-1}}^{K_{j-1}-1} \frac{[\sum_{n\in\mathbb{Z}} d_{nK_{j+1}-K_j-l}(\frac{d_{nK_{j+1}-K_j-l-1}}{d_{nK_{j+1}-K_j-l}} - 1)]^2}{(\sum_{n\in\mathbb{Z}} d_{nK_{j+1}-K_j-l})^2}$$

From Schwartz inequality,

$$I \leq \sum_{l=-K_{j-1}}^{K_{j-1}-1} \frac{\sum_{n\in\mathbb{Z}} d_{nK_{j+1}-K_j-l}^2}{(\sum_{n\in\mathbb{Z}} d_{nK_{j+1}-K_j-l})^2} \cdot \sum_{n\in\mathbb{Z}}\left(\frac{d_{nK_{j+1}-K_j-l-1}}{d_{nK_{j+1}-K_j-l}} - 1\right)^2$$

$$\leq \sum_{l=-K_{j-1}}^{K_{j-1}-1}\sum_{n\in\mathbb{Z}}\left(\frac{d_{nK_{j+1}-K_j-l-1}}{d_{nK_{j+1}-K_j-l}} - 1\right)^2 \leq C,$$

here the last inequality is valid since $\left\{\frac{d_n}{d_{n+1}} - 1\right\} \in l^2$.   □

Now we prove (4.8.1).

From (4.7.4) and Lemma 4.8.2, we have

$$|\tau(L_j)| = \frac{1}{\|L_j\|^2} \cdot \frac{1}{2\pi} \int_0^{2\pi} e^{ix} |L_j(x)|^2 dx$$

$$= \frac{e^{ih_{j+1}}}{\|L_j\|^2 K_j^2} \left[\sum_{l=-K_{j-1}}^{K_{j-1}-1}\sum_{n\in\mathbb{Z}} X_{n,l} + \sum_{n\in\mathbb{Z}}(Y_n + Z_n) \right].$$

It is obvious that $|\tau(L_j)| \leq 1$. From (4.7.3), we obtain

$$\mathrm{Var}(L_j) = 1 - |\tau(L_j)|$$

$$\leq 1 - \frac{1}{\|L_j\|^2 K_j^2} \sum_{l=-K_{j-1}}^{K_{j-1}-1} \sum_{n \in \mathbf{Z}} X_{n,l} \qquad (4.8.17)$$

$$= \frac{1}{\|L_j\|^2 K_j^2} \left| \|L_j\|^2 K_j^2 - \sum_{l=-K_{j-1}}^{K_{j-1}-1} \sum_{n \in \mathbf{Z}} X_{n,l} \right|.$$

Set $R_l = M_l^{j+1} d_{nK_{j+1}-l}$ and $S_l = M_l^{j+1} \tilde{S}_l^{j+1}$. From Lemma 4.8.2 (b) and (4.8.17), we have

$$|\mathrm{Var}(L_j)|$$

$$= \frac{1}{\|L_j\|^2 K_j^2} \left| \sum_{l=-K_{j-1}}^{K_{j-1}-1} \sum_{n \in \mathbf{Z}} \left\{ \frac{R_l^2 + R_{l+K_j}^2}{(S_l + S_{l+K_j})^2} - \frac{R_l R_{l+1} + R_{K_j+l} R_{K_j+l+1}}{[S_l + S_{K_j+l}][S_{l+1} + S_{K_j+l+1}]} \right\} \right|$$

$$\leq \frac{1}{\|L_j\|^2 K_j^2} \sum_{l=-K_{j-1}}^{K_{j-1}-1} \sum_{n \in \mathbf{Z}} \left\{ \left| \frac{R_l^2}{(S_l + S_{l+K_j})^2} - \frac{R_l R_{l+1}}{[S_l + S_{K_j+l}][S_{l+1} + S_{K_j+l+1}]} \right| \right.$$

$$\left. + \left| \frac{R_{l+K_j}^2}{(S_l + S_{l+K_j})^2} - \frac{R_{l+K_j} R_{l+1+K_j}}{[S_l + S_{l+K_j}][S_{l+1} + S_{l+1+K_j}]} \right| \right\}$$

$$= \frac{1}{\|L_j\|^2 K_j^2}(B_1 + B_2). \qquad (4.8.18)$$

By applying Schwartz's inequality, we can estimate B_1 as follows

$$B_1 = \sum_{l=-K_{j-1}}^{K_{j-1}-1} \sum_{n \in \mathbf{Z}} \left(\frac{R_l}{S_l + S_{l+K_j}} \right) \left| \left(\frac{R_l}{S_l + S_{l+K_j}} - \frac{R_{l+1}}{S_{l+1} + S_{l+1+K_j}} \right) \right|$$

$$\leq \sum_{l=-K_{j-1}}^{K_{j-1}-1} \left\{ \sum_{n \in \mathbf{Z}} \left(\frac{R_l}{S_l + S_{l+K_j}} \right)^2 \right\}^{\frac{1}{2}}$$

$$\times \left\{ \sum_{n \in \mathbf{Z}} \left(\frac{R_l}{S_l + S_{l+K_j}} - \frac{R_{l+1}}{S_{l+1} + S_{l+1+K_j}} \right)^2 \right\}^{\frac{1}{2}}.$$

Using Schwartz inequality again, we have

$$
B_1 \leq \left\{ \sum_{l=-K_{j-1}}^{K_{j-1}-1} \sum_{n\in\mathbb{Z}} \left(\frac{R_l}{S_l + S_{l+K_j}} \right)^2 \right\}^{\frac{1}{2}}
$$
$$
\times \left\{ \sum_{l=-K_{j-1}}^{K_{j-1}-1} \sum_{n\in\mathbb{Z}} \left(\frac{R_l}{S_l + S_{l+K_j}} - \frac{R_{l+1}}{S_{l+1} + S_{l+1+K_j}} \right)^2 \right\}^{\frac{1}{2}}.
$$

By Lemma 4.8.2 (b), then

$$
B_1 \leq \|L_j\|K_j \left\{ \sum_{l=-K_{j-1}}^{K_{j-1}-1} \sum_{n\in\mathbb{Z}} \left(\frac{R_l}{S_l + S_{l+K_j}} - \frac{R_{l+1}}{S_{l+1} + S_{l+1+K_j}} \right)^2 \right\}^{\frac{1}{2}}
$$
$$
=: \|L_j\|K_j \cdot I_1. \tag{4.8.19}
$$

Similarly, we can also estimate B_2 and obtain

$$
B_2 \leq \|L_j\|K_j \cdot \left\{ \sum_{l=-K_{j-1}}^{K_{j-1}-1} \sum_{n\in\mathbb{Z}} \left(\frac{R_{l+K_j}}{S_l + S_{l+K_j}} - \frac{R_{l+1+K_j}}{S_{l+1} + S_{l+1+K_j}} \right)^2 \right\}^{\frac{1}{2}}
$$
$$
= \|L_j\|K_j \cdot I_2. \tag{4.8.20}
$$

We conclude from (4.8.18), (4.8.19) and (4.8.20) that

$$
|\mathrm{Var}(L_j)| \leq \frac{I_1 + I_2}{\|L_j\|K_j}. \tag{4.8.21}
$$

In order to establish (4.8.1), it suffices from (4.8.21) to show that there exists a constant C independent of j such that

$$
I_1 + I_2 \leq C \tag{4.8.22}
$$

and

$$
\frac{1}{\|L_j\|K_j} = O\left(\frac{1}{\sqrt{K_j}} \right) \quad \text{as} \quad j \to \infty. \tag{4.8.23}
$$

We estimate I_1 first. From (4.8.19) we have

$$I_1^2 = \sum_{l=-K_{j-1}}^{K_{j-1}-1} \sum_{n \in \mathbf{Z}} \left| \frac{R_{l+1}}{S_{l+1} + S_{l+1+K_j}} - \frac{M_l^{j+1} d_{nK_{j+1}-l-1}}{S_l + S_{l+K_j}} \right.$$

$$\left. + \frac{M_l^{j+1} d_{nK_{j+1}-l-1}}{S_l + S_{l+K_j}} - \frac{R_l}{S_l + S_{l+K_j}} \right|^2$$

$$\leq 2 \sum_{l=-K_{j-1}}^{K_{j-1}-1} \sum_{n \in \mathbf{Z}} \left| \frac{R_{l+1}}{S_{l+1} + S_{l+1+K_j}} - \frac{M_l^{j+1} d_{nK_{j+1}-l-1}}{S_l + S_{l+K_j}} \right|^2$$

$$+ 2 \sum_{l=-K_{j-1}}^{K_{j-1}-1} \sum_{n \in \mathbf{Z}} \left| \frac{M_l^{j+1} d_{nK_{j+1}-l-1}}{S_l + S_{l+K_j}} - \frac{R_l}{S_l + S_{l+K_j}} \right|^2$$

$$= I_{11}^2 + I_{12}^2. \tag{4.8.23}'$$

We will estimate I_{11}^2 and I_{12}^2 respectively,

$$I_{11}^2 = 2 \sum_{l=-K_{j-1}}^{K_{j-1}-1} \sum_{n \in \mathbf{Z}} d_{nK_{j+1}-l-1}^2 \left| \frac{M_{l+1}^{j+1}}{S_{l+1} + S_{l+1+K_j}} - \frac{M_l^{j+1}}{S_l + S_{l+K_j}} \right|^2$$

$$= 2 \sum_{l=-K_{j-1}}^{K_{j-1}-1} \sum_{n \in \mathbf{Z}} d_{nK_{j+1}-l-1}^2$$

$$\times \left| \frac{M_{l+1}^{j+1}(S_l + S_{l+K_j}) - M_l^{j+1}(S_{l+1} + S_{l+1+K_j})}{(S_{l+1} + S_{l+1+K_j})(S_l + S_{l+K_j})} \right|^2$$

$$\leq 4 \sum_{l=-K_{j-1}}^{K_{j-1}-1} \sum_{n \in \mathbf{Z}} d_{nK_{j+1}-l-1}^2 \left\{ \left[\frac{M_{l+1}^{j+1} S_l - M_l^{j+1} S_{l+1}}{(S_{l+1} + S_{l+1+K_j})(S_l + S_{l+K_j})} \right]^2 \right.$$

$$\left. + \left[\frac{M_{l+1}^{j+1} S_{l+K_j} - M_l^{j+1} S_{l+1+K_j}}{(S_{l+1} + S_{l+1+K_j})(S_l + S_{l+K_j})} \right]^2 \right\}$$

$$\leq 4 \sum_{l=-K_{j-1}}^{K_{j-1}-1} \sum_{n \in \mathbf{Z}} d_{nK_{j+1}-l-1}^2 \left[\frac{M_{l+1}^{j+1} S_l - M_l^{j+1} S_{l+1}}{S_{l+1} S_l} \right]^2$$

$$+4 \sum_{l=-K_{j-1}}^{K_{j-1}-1} (\tilde{S}_{l+1}^{j+1})^2 \left| \frac{(M_{l+1}^{j+1})^2 \tilde{S}_{l+K_j}^{j+1} - (M_l^{j+1})^2 \tilde{S}_{l+1+K_j}^{j+1}}{[(M_{l+1}^{j+1})^2 \tilde{S}_{l+1}^{j+1} + \tilde{S}_{l+1+K_j}^{j+1}][(M_l^{j+1})^2 \tilde{S}_l^{j+1} + \tilde{S}_{l+K_j}^{j+1}]} \right|^2$$

$$\leq 4 \sum_{l=-K_{j-1}}^{K_{j-1}-1} \sum_{n \in \mathbf{Z}} d_{nK_{j+1}-l-1}^2 \left| \frac{1}{\tilde{S}_{l+1}^{j+1}} - \frac{1}{\tilde{S}_l^{j+1}} \right|^2$$

$$+4 \sum_{l=-K_{j-1}}^{K_{j-1}-1} (\tilde{S}_{l+1}^{j+1})^2 \left[\frac{(M_{l+1}^{j+1})^2 \tilde{S}_{l+K_j}^{j+1} - (M_l^{j+1})^2 \tilde{S}_{l+1+K_j}^{j+1}}{(M_{l+1}^{j+1})^2 \tilde{S}_{l+1}^{j+1} [(M_l^{j+1})^2 \tilde{S}_l^{j+1} + \tilde{S}_{l+K_j}^{j+1}]} \right]^2 . \quad (4.8.24)$$

The first term in the right hand side of (4.8.24) is

$$4 \sum_{l=-K_{j-1}}^{K_{j-1}-1} \sum_{n \in \mathbf{Z}} d_{nK_{j+1}-l-1}^2 \left| \frac{1}{\tilde{S}_{l+1}^{j+1}} - \frac{1}{\tilde{S}_l^{j+1}} \right|^2$$

$$\leq 8 \sum_{l=-K_{j-1}}^{K_{j-1}-1} \sum_{n \in \mathbf{Z}} \left\{ \left| \frac{d_{nK_{j+1}-l-1}}{\tilde{S}_{l+1}^{j+1}} - \frac{d_{nK_{j+1}-l}}{\tilde{S}_{l+1}^{j+1}} \right|^2 + \left| \frac{d_{nK_{j+1}-l} - d_{nK_{j+1}-l-1}}{\sum_{m \in \mathbf{Z}} d_{mK_{j+1}-l}} \right|^2 \right\}$$

$$\leq 8 \left\{ M_0 + \sum_{l=-K_{j-1}}^{K_{j-1}-1} \sum_{n \in \mathbf{Z}} \left| 1 - \frac{d_{nK_{j+1}-l-1}}{d_{nK_{j+1}-l}} \right|^2 \right\}$$

$$\leq 8 \{ M_0 + C \}, \quad (4.8.25)$$

where the constant M_0 comes from the estimate (4.7.9); and the constant C comes from the assumption that $\left\{ \dfrac{d_{n+1}}{d_n} - 1 \right\} \in l^2$.

Denote the second term in the right hand side of (4.8.24) by

$$\eta = 4 \sum_{l=-K_{j-1}}^{K_{j-1}-1} (\tilde{S}_{l+1}^{j+1})^2 \left[\frac{(M_{l+1}^{j+1})^2 \tilde{S}_{l+K_j}^{j+1} - (M_l^{j+1})^2 \tilde{S}_{l+1+K_j}^{j+1}}{(M_{l+1}^{j+1})^2 \tilde{S}_{l+1}^{j+1} [(M_l^{j+1})^2 \tilde{S}_l^{j+1} + \tilde{S}_{l+K_j}^{j+1}]} \right]^2 .$$

Divide the numerator and the denominator by $(M_{l+1}^{j+1})^2$, then

$$\eta = \sum_{l=-K_{j-1}}^{K_{j-1}-1} \left[\frac{(\frac{M_l^{j+1}}{M_{l+1}^{j+1}})^2 \tilde{S}_{l+1+K_j}^{j+1} - \tilde{S}_{l+K_j}^{j+1}}{(M_l^{j+1})^2 \tilde{S}_l^{j+1} + \tilde{S}_{l+K_j}^{j+1}} \right]^2$$

$$\leq 2 \sum_{l=-K_{j-1}}^{K_{j-1}-1} \left\{ \left[\frac{[(\frac{M_l^{j+1}}{M_{l+1}^{j+1}})^2 - 1]\tilde{S}_{l+1+K_j}^{j+1}}{(M_l^{j+1})^2 \tilde{S}_l^{j+1} + \tilde{S}_{l+K_j}^{j+1}} \right]^2 + \left[\frac{\tilde{S}_{l+1+K_j}^{j+1} - \tilde{S}_{l+K_j}^{j+1}}{(M_l^{j+1})^2 \tilde{S}_l^{j+1} + \tilde{S}_{l+K_j}^{j+1}} \right]^2 \right\}$$

$$\leq 2 \sum_{l=-K_{j-1}}^{K_{j-1}-1} \left\{ \left| \left(\frac{M_l^{j+1}}{M_{l+1}^{j+1}} \right)^2 - 1 \right|^2 \left| \frac{\tilde{S}_{l+1+K_j}^{j+1} - \tilde{S}_{l+K_j}^{j+1} + \tilde{S}_{l+K_j}^{j+1}}{(M_l^{j+1})^2 \tilde{S}_l^{j+1} + \tilde{S}_{l+K_j}^{j+1}} \right|^2 \right\}$$

$$+2 \sum_{l=-K_{j-1}}^{K_{j-1}-1} \left| \frac{\tilde{S}_{l+1+K_j}^{j+1}}{\tilde{S}_{l+K_j}^{j+1}} - 1 \right|^2 .$$

We use the inequality $|a + b|^2 \leq 2(|a|^2 + |b|^2)$ again to obtain

$$\eta \leq 2 \sum_{l=-K_{j-1}}^{K_{j-1}-1} \left| \frac{\tilde{S}_{l+1+K_j}^{j+1}}{\tilde{S}_{l+K_j}^{j+1}} - 1 \right|^2 + 4 \sum_{l=-K_{j-1}}^{K_{j-1}-1} \left| \left(\frac{M_l^{j+1}}{M_{l+1}^{j+1}} \right)^2 - 1 \right|^2$$

$$\times \left[\left| \frac{\tilde{S}_{l+1+K_j}^{j+1} - \tilde{S}_{l+K_j}^{j+1}}{\tilde{S}_{l+K_j}^{j+1}} \right|^2 + \left| \frac{\tilde{S}_{l+K_j}^{j+1}}{(M_l^{j+1})^2 \tilde{S}_l^{j+1} + \tilde{S}_{l+K_j}^{j+1}} \right|^2 \right]$$

$$\leq 4 \sum_{l=-K_{j-1}}^{K_{j-1}-1} \left| \left(\frac{M_l^{j+1}}{M_{l+1}^{j+1}} \right)^2 - 1 \right|^2 \left\{ \left| \frac{\tilde{S}_{l+1+K_j}^{j+1}}{\tilde{S}_{l+K_j}^{j+1}} - 1 \right|^2 + 1 \right\}$$

$$+2 \sum_{l=-K_{j-1}}^{K_{j-1}-1} \left| \frac{\tilde{S}_{l+1+K_j}^{j+1}}{\tilde{S}_{l+K_j}^{j+1}} - 1 \right|^2 .$$

By Schwartz' inequality, (4.8.11), and (4.8.16), we have

$$\eta \leq 4 \left(\sum_{l=-K_{j-1}}^{K_{j-1}-1} \left| \left(\frac{M_l^{j+1}}{M_{l+1}^{j+1}} \right)^2 - 1 \right|^4 \right)^{\frac{1}{2}} \left(\sum_{l=-K_{j-1}}^{K_{j-1}-1} \left| \frac{\tilde{S}_{l+1+K_j}^{j+1}}{\tilde{S}_{l+K_j}^{j+1}} - 1 \right|^4 \right)^{\frac{1}{2}} + C$$

$$\leq 4 \left(\sum_{l=-K_{j-1}}^{K_{j-1}-1} \left| \left(\frac{M_l^{j+1}}{M_{l+1}^{j+1}} \right)^2 - 1 \right|^2 \right) \left(\sum_{l=-K_{j-1}}^{K_{j-1}-1} \left| \frac{\tilde{S}_{l+1+K_j}^{j+1}}{\tilde{S}_{l+K_j}^{j+1}} - 1 \right|^2 \right) + C$$

$$\leq C. \tag{4.8.26}$$

As for the second term I_{12}^2 in $(4.8.23)'$, it can be estimated as follows

$$I_{12}^2 = 2 \sum_{l=-K_{j-1}}^{K_{j-1}-1} \sum_{n \in \mathbb{Z}} \left(\frac{M_l^{j+1} d_{nK_{j+1}-l}}{M_l^{j+1} \tilde{S}_l^{j+1} + (M_l^{j+1})^{-1} \tilde{S}_{l+K_j}^{j+1}} \right)^2 \left(\frac{d_{nK_{j+1}-l-1}}{d_{nK_{j+1}-l}} - 1 \right)^2$$

$$\leq 2 \sum_{l=-K_{j-1}}^{K_{j-1}-1} \sum_{n \in \mathbb{Z}} \left(\frac{M_l^{j+1} d_{nK_{j+1}-l}}{M_l^{j+1} \tilde{S}_l^{j+1}} \right)^2 \left(\frac{d_{nK_{j+1}-l-1}}{d_{nK_{j+1}-l}} - 1 \right)^2$$

$$\leq 2 \sum_{l=-K_{j-1}}^{K_{j-1}-1} \sum_{n \in \mathbb{Z}} \left(\frac{d_{nK_{j+1}-l-1}}{d_{nK_{j+1}-l}} - 1 \right)^2 \leq C. \tag{4.8.27}$$

From $(4.8.23)'$, $(4.8.25)$, $(4.8.26)$, and $(4.8.27)$, we conclude that

$$I_1 \leq C \tag{4.8.28}$$

for some constant C.

Replacing $(4.8.11)$ by $(4.8.12)$ in above procedure, we obtain

$$I_2 \leq C, \tag{4.8.29}$$

thus $(4.8.22)$ is proved.

In order to establish $(4.8.23)$, we notice that

$$K_j^2 \|L_j\|^2 = \sum_{l=-K_{j-1}}^{K_{j-1}-1} \frac{(M_l^{j+1})^4 Q_l + Q_{l+K_j}}{[(M_l^{j+1})^2 \tilde{S}_l^{j+1} + \tilde{S}_{l+K_j}^{j+1}]^2}$$

$$\geq \frac{1}{2} \sum_{l=-K_{j-1}}^{K_{j-1}-1} \frac{(M_l^{j+1})^4 Q_l + Q_{l+K_j}}{[(M_l^{j+1})^2 \tilde{S}_l^{j+1}]^2 + (\tilde{S}_{l+K_j}^{j+1})^2}$$

$$= \frac{1}{2} \sum_{l=-K_{j-1}}^{K_{j-1}-1} \frac{Q_{l+K_j} + Q_l}{\frac{Q_{l+K_j}}{Q_l}(\tilde{S}_l^{j+1})^2 + \frac{Q_l}{Q_{l+K_j}}(\tilde{S}_{l+K_j}^{j+1})^2}. \tag{4.8.30}$$

Now we estimate the denominator in $(4.8.30)$. The condition $(4.7.17)$ leads to the following inequality:

$$\frac{Q_l}{(\tilde{S}_l^{j+1})^2} \geq \left(\frac{d_{-l}}{\tilde{S}_l^{j+1}} \right)^2 \geq C^2, \quad \text{for} \quad -K_j \leq l \leq K_j,$$

in turn,

$$(\tilde{S}_l^{j+1})^2 \le \frac{Q_l}{C^2} \quad \text{for} \quad -K_{j-1} \le l \le K_{j-1} - 1. \tag{4.8.31}$$

It is easily seen that

$$\frac{(\tilde{S}_{l+K_j}^{j+1})^2}{Q_{l+K_j}} = \frac{(\tilde{S}_{l-K_j}^{j+1})^2}{Q_{l-K_j}}.$$

Since

$$l - K_j \in \{-K_j, -K_{j+1}, \cdots, K_j\} \quad \text{if} \quad l \in \{0, 1, \cdots, K_{j-1} - 1\}$$

by (4.7.17) we obtain

$$\frac{d_{-(l-K_j)}}{\tilde{S}_{l-K_j}^{j+1}} \ge C,$$

which implies that

$$\begin{aligned}
\frac{(\tilde{S}_{l-K_j}^{j+1})^2}{Q_{l-K_j}} &= \frac{(\tilde{S}_{l-K_j}^{j+1})^2}{d_{-(l-K_j)}^2 + \sum_{n \neq 0} d_{nK_{j+1}-(l-K_j)}} \\
&\le \frac{(\tilde{S}_{l-K_j}^{j+1})^2}{d_{-(l-K_j)}^2} \\
&\le \frac{1}{C^2}. \tag{4.8.32}
\end{aligned}$$

On the other hand

$$K_j + l \in \{-K_j, -K_j + 1, \cdots, K_j\} \quad \text{if} \quad l \in \{-K_{j-1}, \cdots, -1\},$$

hence we also have

$$\frac{(\tilde{S}_{l+K_j}^{j+1})^2}{Q_{l+K_j}} \le \frac{1}{C^2}. \tag{4.8.33}$$

Therefore, from (4.8.32) and (4.8.33)

$$(\tilde{S}_{l+K_j}^{j+1})^2 \le \frac{1}{C^2} Q_{l+K_j} \quad \text{for} \quad l \in \{-K_{j-1}, \cdots, K_{j-1} - 1\}. \tag{4.8.34}$$

From (4.8.30), (4.8.31) and (4.8.34)

$$\begin{aligned}
K_j^2 \|L_j\|^2 &\ge \frac{1}{2} \sum_{l=-K_{j-1}}^{K_{j-1}-1} \frac{(Q_{l+K_j} + Q_l)C^2}{Q_{l+K_j} + Q_l} \\
&= \frac{C^2}{2} K_j. \tag{4.8.35}
\end{aligned}$$

On the other hand, by the inequality $Q_l \leq (\tilde{S}_l^{j+1})^2$,

$$K_j^2 \|L_j\|^2 \leq \sum_{l=-K_{j-1}}^{K_{j-1}-1} \left\{ \frac{Q_l}{(\tilde{S}_l^{j+1})^2} + \frac{Q_{l+K_j}}{(\tilde{S}_{l+K_j}^{j+1})^2} \right\} \tag{4.8.36}$$

$$\leq 2K_j.$$

now (4.8.35) and (4.8.36) together yield

$$\frac{C^2}{2K_j} \leq \|L_j\|^2 \leq \frac{2}{K_j}. \tag{4.8.37}$$

Thus (4.8.23) follows.

Using (4.8.23), (4.8.22) and (4.8.21) we conclude that

$$|\mathrm{Var}(L_j)| = O\left(\frac{1}{\sqrt{K_j}}\right);$$

this is (4.8.1).   □

As indicated above, we have proved the local property of the PCIW. We now ask of their dual functions: do the Dual scaling function and the dual wavelet also enjoy the local property?

§4.9   Examples

In this section we shall give some examples of PISF and PCIW.

Define a function

$$P_{2n}(t) = 1 + \sum_{\nu \neq 0} \frac{e^{i\nu t}}{\nu^{2n}}. \tag{4.9.1}$$

This is a special case of the function ϕ_m defined in (4.6.18). In fact, $P_{2n}(t) = (-1)^{n+1}\phi_{2n}(t)$ with $m = 2n, T = 2\pi$. Now, the generator is

$$g(t) = P_{2n}(t) = \sum_{\nu \in \mathbb{Z}} d_\nu e^{i\nu t},$$

where

$$d_k = \begin{cases} 1, & k = 0 \\ \frac{1}{k^{2n}}, & k \neq 0 \end{cases}$$

We have to check the following requirements:

$$\frac{d_{k+1}}{d_k} - 1 \in l^2 \tag{4.9.2}$$

and

$$\inf_{\substack{|l| \le K_{j-1} \\ j \ge 0}} \frac{d_{-l}}{\tilde{S}_l^j} \ge C > 0 \tag{4.9.3}$$

(see (4.7.18), (4.7.17) and Theorem 4.8.1).

Set

$$a_k = \left(\frac{d_{k+1}}{d_k} - 1\right)^2,$$

then

$$a_k = \left(\frac{1}{k+1}\right)^2 \left[4n^2 + O\left(\frac{1}{k+1}\right)\right].$$

It is clear that $\{a_k\}_{k \in \mathbb{Z}} \in l^2$.

Set

$$F_{l,j} = \frac{d_{-l}}{\tilde{S}^j}.$$

Then

$$F_{l,j} = F_{-l,j}$$

$$= \left\{\sum_{\nu \ne 0} \frac{|l|^{2n}}{(\nu K_j + |l|)^{2n}} + 1\right\}^{-1}, \quad |l| \le K_{j-1} \tag{4.9.4}$$

$$F_{0,j} = \left\{\sum_{\nu \ne 0} \frac{1}{(\nu K_j)^{2n}} + 1\right\}^{-1}. \tag{4.9.5}$$

It is easy to show

$$\inf_{\substack{|l| \le K_{j-1} \\ j \ge 0}} F_{l,j} = \min_{j \ge 0}(F_{K_{j-1},j}, F_{0,j})$$

$$= \min_{j \ge 0}\left\{\left[\sum_{\nu \ne 0} \frac{1}{(2\nu + 1)^{2n}} + 1\right]^{-1}, \left[\sum_{\nu \ne 0} \frac{1}{(\nu K_j)^{2n}} + 1\right]^{-1}\right\}$$

$$= \min\left\{\left[\sum_{\nu \ne 0} \frac{1}{(2\nu + 1)^{2n}} + 1\right]^{-1}, \left[\sum_{\nu \ne 0} \frac{1}{(\nu K)^{2n}} + 1\right]^{-1}\right\}$$

$$> 0.$$

Thus for the generator defined in (4.9.1), the corresponding coefficients satisfy the requirements (4.9.2) and (4.9.3).

We point out that $P_{2n}(t)$ is a polynomial of degree $2n$ in one period. For $n = 1$ and $n = 2$, they have the following form:

$$P_2(t) = \frac{6 - \pi^2}{6} + \frac{1}{2}(t - \pi)^2, \tag{4.9.6}$$

and

$$P_4(t) = \frac{360 - 7\pi^4}{360} + \frac{\pi^2(t - \pi)^2}{12} - \frac{(t - \pi)^4}{24}, \tag{4.9.7}$$

where $0 \le t \le 2\pi$ and we assume $K = 1$.

In the following we plot the graphs of the scaling functions and the wavelets generated by $P_2(t)$ and $P_4(t)$, respectively.

In the first place the generator is assumed to be $P_2(t)$.

Fig. 1 is the graph of the scaling function $\phi_4(t)$ with 16 knots (as in (4.1.8); ϕ_j, where $j = 4$, has $2^4 (= 16)$ knots).

Fig. 2 is the graph of corresponding $L_4(t)$.

Fig. 5 and Fig. 6 are the graphs of ϕ_5 and L_5 respectively; they all have $32 (= 3^5, j = 5)$ knots.

These are in $C^0([0, 2\pi])$.

In the second case the generator is $P_4(t)$. The corresponding scaling functions and wavelets are $\phi_4(t), \phi_5(t), L_4(t)$ and $L_5(t)$.

Fig. 3 is the graph of $\phi_4(t)$ with 16 knots.

Fig. 4 is the corresponding wavelet function $L_4(t)$.

Fig. 7 and Fig. 8 are the graphs for $\phi_5(t)$ and L_5 with 32 knots, respectively.

These functions are in $C^2([0, 2\pi])$.

The graphs show that ϕ_j and L_j are symmetric. We can also see that the localization behaves better if j gets larger.

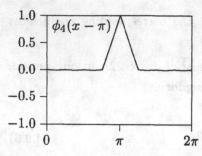

FIG 1. $\phi_4(x - \pi)$ generated by $P_2(t)$

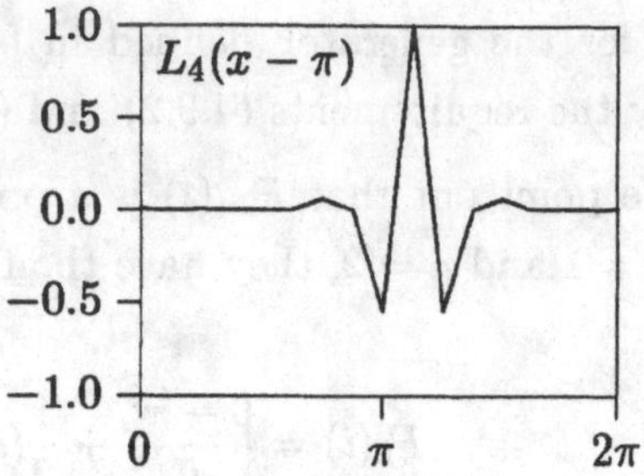

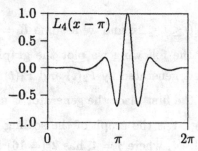

FIG 2. L_4 generated by $P_2(t)$

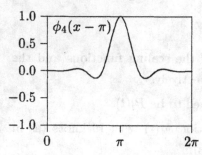

FIG 3. $\phi_4(x - \pi)$ generated by $P_4(t)$

FIG 4. $L_4(x - \pi)$ generated by $P_4(t)$

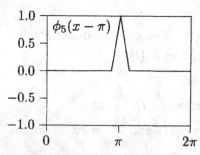

FIG 5. $\phi_5(x - \pi)$ generated by $P_2(t)$

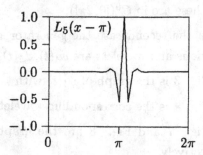

FIG 6. $L_5(x - \pi)$ generated by $P_2(t)$

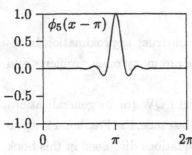
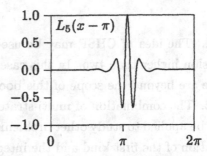

FIG 7. $\phi_5(x - \pi)$ generated by $P_4(t)$ FIG 8. $L_5(x - \pi)$ generated by $P_4(t)$

Note 1.   The proof of these assertions is analogous to that of Theorem 2.4.1.

Note 2.   Lemma 4.6.1 was established by Schoenberg [Sch2], Lemma 4.6.2 was proved by Micchelli for the L-spline case [M].

Note 3.   The proof of the existence and uniqueness of the interpolatory spline function with mixed boundary conditions was given by Karlin and Pinkus [KP]. The fundamental theorem of algebra in the more general case was studied by Chen Han-lin (see [C6]).

Note 4.   Set $T = 2\pi$ and $\phi_n^*(x) = -\frac{1}{2}\phi_n(x)$. Some authors call $\phi_n^*(x)$ the Bernoulli polynomial on $[0, 2\pi]$ (see [BHS]).

Note 5.   The proof can be found in Meinardus' work [Mei] and in [BHS].

Concluding Remarks

1. The idea of CHSF may be used to construct approximation in dimension higher than two. In this case there are many open problems and there are beyond the scope of this book.

2. The combination of multi-strategy and PQW (or its generalization) may be applied to study other problems; for instance, the Fredholm integral equation of the first kind and the integral equations discussed in this book on higher-dimensional manifolds.

REFERENCES

[Ah] Ahlberg J. H., *Splines in the complex plane, in: Approxima-tions with special emphasis on spline functions*, I. J. Schoen-berg, Ed. 1–27, Academic Press, New York, 1969.

[ANW1] ——, Nilson, E. N., Walsh, J. L., Complex polynomial splines on the unit circle, *J. Math. Analysis and Appl.* **33**, 2(1971), pp. 234–257.

[ANW2] ——, Nilson E. N., Walsh, J. L., Complex cubic splines, *Trans. Amer. Math. Soc.* **129**(1967), pp. 321–413.

[BB] Brewster M. E. & Beylkin G., A multiresolution strategy for numerical homogenization, *Appl. Comp. Anal.*, **2**(1995), pp. 327–349.

[BCR] Beylkin G., Coiffman R. R., & Rokhlin V., Fast wavelet trans-forms and numerical algorithms I, *Comm. Pure Appl. Math.* **44**(1991), pp. 141–183.

[BF] C. de Boor & Fix G. J., Spline approximation by quasiiter-polants, *J. Approx. Theory*, **8**(1973), pp. 19–45.

[BHS] Bojanov B. D., Hakopian H. A. & Sahakian A. A., *Spline functions and multivariate interpolations*, Kluwer Academic Publishers. Dordrecht/Boston/London, 1993.

[C1] Chen, H. L., Quasiinterpolant spline on the unit circle, *J. Ap-prox. Theory*, **38**, 4(1983), pp. 312–318.

[C2] ——, Wavelets on the unit circle, *Result in Mathemat-ics*, **31**(1997), pp. 322–336.

[C3] ——, Complex spline functions, *Scientia Sinica*, **24**, 2(1981), pp. 160–169.

[C4] ——, Interpolation and approximation on the unit disc by complex harmonic splines, *J. Approx. Theory*, **43**, 2(1985), pp. 112–123.

[C5] ——, The zeros of rational splines and complex splines, *J. Approx. Theory*, **39**, 4(1983), pp. 308–319.

213

[C6] ————, The zeros of G-spline and interpolation by G-splines with mixed boundary conditions, *Approx. Theory and its Applications*, **1**, 2(1985), pp. 1–14.

[C7] ————, Interpolation by splines on finite and infinite planar sets, *Chin. Ann. Math.* **5B**, 3(1984), pp. 375–390.

[C8] ————, Antiperiodic wavelets, *J. Comp. Math.* **14**, 1(1996), pp. 32–39.

[C9] ————, Wavelets from trigonometric spline approach, *Approx. Theory and its Applications*, **12**, 2(1996), pp. 99–110.

[C10] ————, Periodic orthonormal quasi-wavelet bases, *Chinese Science Bulletin*, **41**, 7(1996), pp. 552–554.

[C11] ————, Properties of complex harmonic splines, *Numer. Funct. Anal. and Optimiz.*, **13**(3 & 4)(1992), pp. 233–242.

[C12] ————, Typical real functions, *J. Fudan Univ.* **1.3**, 3(1957).

[C13] ————, Extremal quasiconformal mapping on hexagon, *Progress in Math. Sinica* **1**, 3(1960).

[C14] ————, The extremal quasiconformal mapping on heptagon, *Progress in Math. Sinica* **5**, 4(1962).

[C15] ————, Quasiconformal mapping on N-dimensional space, *Acta Math. Sinica* bf 14, 11(1964).

[CC] —— & Chui, C. K.   On a generalized Euler spline and its application to the study of convergence in cardinal interpolation and solution of extremal problems, *Acta Math. Hung.* **61** (3–4)(1993), pp. 219–233.

[CH1] —— & Hvaring T.,   Approximation of complex harmonic functions by complex harmonic splines, *Math. Comp.* **42**, 165(1984), pp. 151–164.

[CH2] ——, A new method for the approximation of conformal mapping on the unit circle. *Tech. Report, Math. and Comp.* 6/82, ISBN 82-7151-049-5.

[CLJ] ——, Liang X. Z. & Jin, G. R.   Bivariate box-spline wavelets, in:*Harmonic Analysis in China* (M. T. Cheng etc. eds), pp. 183–196, Kluwer Academic Publishers, 1995.

[CLPX] ——, Liang, X. Z., & Peng, S. L., Real valued periodic wavelets: construction and the relation with Fourier series, to be appeared in *J. Comp. Math.*

[CLdP] ——, Li, D. F., & Peng, S. L., Periodic interpolatory wavelets, to appear.

[CP1] ——, Peng, S. L., Solving integral equations with logarithmic kernel by using periodic quasi-wavelet, to be published in *J. Comp. Math.*.

[CP2] ——, Peng, S. L., An $O(N)$ quasi-wavelet algorithm for a second kind boundary integral equation with a logarithmic kernel. to appear.

[CP3] ——, Peng, S. L., Local Properties of cardinal interpolatory function. to appear in *ACTA mathematica Sinica*.

[CX] ——, Xiao, S. L., Periodic cardinal interpolatory wavelets, *Chin. Ann. of Math.* **198**, 2(1998), pp. 133-142.

[Ct] Chen, T. P. Complex splines and an extremal problem, *Chinese Science Bulletin* **32**, 1(1987), pp. 1–8.

[CMX] Chen, Z., Micchelli C. A. & Xu, Y., The Petrov-Galerkin methods for second integral equations II: Multiwavelet scheme, *Advances in Computational Mathematics*, **7**(1997), pp. 199–233.

[CM] Chui, C. K. & Mhashar H. N., On trigonometric wavelets, *Const. Approx.* **9**(1993), pp. 167–190.

[Da] Daubechies I., *Ten Lectures on Wavelets*, SIAM, Philadelphia, 1992.

[Dav] Davis P. J., *Introduction and Approximation*, Blaisdell, New York, 1963.

[DKPS] Dahmen W., Kleemann B., Proessdorf S. & Schneider R. Multiscale methods for the solution of the Helmholtz and Laplace equations, preprint.

[DG] Daier D. *Konstruktive Methoden der konformen Abbildung.* Springer Tracts in Natural Philosophy, **3**, 1964.

[Ga] Gantmacher F. R., *The Theory of Matrices*, N. Y. Chelsea

Publ. Co. 1959.

[Gi] Gibbs W. J., *Conformal Transformations in Electrical Engineering*. The British Thomson-Houston Co. LTD., 1958.

[GL1] Goodman T. N. T. & Lee S. L., *B*-splines on the circle and trigonometric *B*-splines, in: *Approx. Th. and Spline Functions* S. P. Singh et al. (eds), D. Reidel Publishing Company, (1984), pp. 297–325.

[GL2] —— & ——, Interpolatory and variation - diminishing properties of generalized *B*-splines, *Proceedings of the Royal Society of Edinburgh*, **96A**(1984), pp. 249–259.

[GLS] —— ——, & Sharma A. Approximation by Λ-splines on the circle, *Can. J. Math.* **XXXVII,** 6(1985), pp. 1085–1111.

[GO] Golusin G. M., *Geometrische Funktionentheorie, VEB Deutscher Verlag der Wissenschaften,* Berlin, 1957.

[GW] Greenspan D. & Werner P., A Numerical method for the exterior Dirichlet problem for the reduced waved equation, *Arch. Rational Mech. Anal.,* 23(1966), pp. 288–316.

[H] Henrici P. *Applied and Computational Complex Analysis,* **3**, Pure & Applied Mathematics, A Wiley-Interscience Publication, 1977.

[KTM] Kamada M., Toraichi K. & Mori R., Periodic orthonormal bases, *J. Approx. Th.* **55**(1988), pp. 27–34.

[KP] Karlin S. & Pinkus A. Interpolation by splines with mixed boundary conditions, in:*Studies in Spline Functions and Approximation Theory,* Acad. Press Inc., New York, 1976.

[Ke] Kelley, C. T., A fast multilevel algorithm for integral equations, *SIAM J. Numer. Anal.,* **32**, 2(1995), pp. 501–513.

[KLT] Koh Y. W., Lee S. L. Tan H. H., Periodic orthogonal splines and wavelets, *Applied and computational harmonic analysis,* **2**(1995), pp. 201–218.

[Kr] Kress R., *Linear Integral Equations,* Springer-Verlag, Berlin, Heidelberg, 1989.

[KS1] Kress R. & Spassov W. T., On the condition number of boundary integral operators for exterior Dirichlet problem for the Helmholtz equation, *Numer. Math.*, **42**(1988), pp. 77–95.

[KS2] Kress R. & Sloan L. H., On the numerical solution of a logarithmic integral equation of the first kind for the Helmholtz equation, *Numer. Math.*, **66**(1993), pp. 199–214.

[Kz] Künzi H. P., *Quasikonforme Abbildungen*, Springer-Verlag, Berlin Göttingen Heidelberg, 1960.

[LS] Lavrentev M. A. & Shabat B., *The Methods in Theory of Complex Functions*, Chinese translation from Russian, (1951) Chapt. 2 Sec. 3.

[LP] Li, D. F. & Peng, S. L., Characterization of periodic multiresolution analysis and an application, *ACTA Mathematica Sinica*, **44**, 4(1998), pp. 547–554.

[Lu] Lu, C. K. Error analysis for interpolating complex cubic splines with deficiency 2, *J. Approx. Theory*, **36**, 3(1982), pp. 183–196.

[Mars] Marsden M. J., *J. Approx. Theory*, 3(1970), pp. 7–49.

[Mar] Markyshevitch A., *Theory of Analytic Functions*, Chinese translation from Russian, (1950) Chapt. 5, Sec. 3.

[MaS] Mathur K. K. & Scharma A., Discrete polynomial splines on the circle, *Acta Math. Hung.* **33**, (1–2)(1979), pp. 143–153.

[Mei] Meinardus, G., Periodische Spline functionen, in: *Spline Functions, Karlsruhe 1975*(K. Böhmer, G. Meinardus, and W. Schempp, Eds.), Lecture Notes in Mathematics Ser., **501**, pp. 177–199, Springer-Verlag, Berlin, 1976.

[Mi] Micchelli C. A., Cardinal *L*-splines, in: *Studies in Spline Functions and Approximation Theory*(S. Karlin, C. A. Micchelli, A. Pinkus and I. J. Schoenberg, eds) pp. 203–250, Academic Press, New York, 1996.

[MS] Micchelli C. A. & Sharma A., Spline functions on the circle: Cardinal *L*-splines revisited, *Cand. J. Math.*, **32**(1980), pp. 1459–1473.

[Mu] Muskhelishvili, N. I., *Singular Integral Equations*, translation from Russian (J.R. M. Radok edited), P. Noordhoff N. V. Groningen-Holland, 1953.

[My]   Meyer Y.,   *Ondelettes et fonctions splines*, Seminaire EDP, Paris, 1996.

[Myr] Meyer Y.,   *Ondelettes et operateurs*, Herman, Paris, 1990.

[Ne]   Nehai Zeev,   *Conformal Mapping*, McGraw-Hill, New York, 1952.

[NW]   Narcowich F. J. & Ward J. D.,   Wavelets associated with periodic basis functions, *Appl. Comput. Harmonic Anal.*, **3**(1996), pp. 40–56.

[PB]   Perrier V. & Basdevant C.,   La Decompasition on Ondelettes Periodiques, un Outil Pour L'analyse de Champs Inhomogenes. Theorie et Algorithmes, *La Recherche Aerospatial*, **3**(1989), pp. 53–67.

[PT1] Plonka G. & Tasche M.,   A unified approach to periodic wavelets, in: *Wavelets: Theory, Algorithms and Applications*, (C.K. Chui, L. Montefusco and Puccis eds.) Academic Press, San Diego, 1994, pp. 137–151.

[PT2] ————,   Periodic spline wavelets, *Technical Report*, **93/94**, FB Mathematik, Universitat Rostock, Germany, 1993.

[PT3] ————,   On the computation of periodic wavelets, *Appl. Comput. Harmonic Anal.*, **2**(1995), pp. 1–14.

[R1]   Reichel L.,   On polynomial approximation in the complex plane with application to conformal mapping, *TRITA-Na-8102*, Dept. Comput. Sci., Royal Institute of Technology, Stockholm.

[R2]   ————,   On the determination of boundary collocation points for solving some problems for the Laplace operator, *TRITA-NA-8006*, Dept. Comput. Sci., Royal Institute of Technology, Stockholm.

[R]   Rudin W.,   *Real and Complex Analysis*, WcGraw-Hill, Inc., New York, 1987.

[Sch1] Schoenberg I. J.,   On polynomial spline functions on the circle (I and II), in: *Proceedings of the conference on constructive theory of functions* (G. Alexits and S. B. Steckin eds.) Budapest, 1972, pp. 403–433.

[Sch2] ————, Cardinal Interpolatory and Spline Functions: II Interpolatory of Data of Power Growth, *J. Approx. Th.* **6**(1972), pp. 404–420.

[Schm] Schmidt G., On complex spline interpolation on the unit circle, in: *Constructive Theory of Functions 1984, Sofia*, pp. 803–807, (1984).

[Sc] Schumaker L. L., *Spline Functions, Basic Theory*, A Wiley-Interscience Publication, 1981.

[SL] Schinzinger R. & Laura P. A. A., *Conformal Mapping: Methods and Applications*, Elsevier Science Publishers, 1991.

[Sm] Smithies F., *Integral Equations*, Cambridge University Press, 1958.

[T] Tsuji M., *Potential Theory in Modern Function Theory*, Maruzen, Tokyo, 1959.

[Tz] Tzimbalario J., Interpolation by complex splines, *Trans. Amer. Math. Soc.*, **243**(1978), pp. 213–222.

[W] Walz G., *Spline Funktionen im Komplexen*, Wissenschaftsverlag, 1991.

[Wr1] Wronicz Z., *Approximation by complex splines*, Zeszgty naukowe uniwersytetu jagiellońsbiego, Prace Matematyczne, Z. 20, (1979).

[Wr2] ————, *On the application of the orthonormal Franklin system to the approximation of analytic functions*, Approximation theory, Banach center publication, **4**, PWN-Polish Scientific Publishers, Warsaw (1979).

[Xu] Xu, S. Y., Complex cubic spline interpolation, *J. College Comp. Math.*, **2**(1982).

[[XZ]] Yuesheng Xu & Yunhe Zhao, An Extrapolation method for a class of boundary integral equations, *Math. Comp.* **65**, 214(1996), pp. 587–610.

[Ya1] Yan Y., A fast boundary element method for the two dimensional Helmholtz equations, *Comput. Method. Appl. Mech.*

220

Engrg., to appear.

[Ya2] ————, A fast numerical solution for a second kind boundary integral equation with a logarithmic kernel, *SIAM J. Numer. Anal.*, **31**, 2(1994), pp. 477–498.

[Zh] Zhu C. Q., Two-dimensional periodic cardinal interpolatory wavelets, to appear.

Index

Author Index